The Politics of Policy in Boys' Education

The Politics of Policy in Boys' Education

Getting Boys "Right"

Marcus B. Weaver-Hightower

THE POLITICS OF POLICY IN BOYS' EDUCATION
Copyright © Marcus B. Weaver-Hightower, 2008.
Softcover reprint of the hardcover 1st edition 2008 978-0-230-60839-9
All rights reserved.

First published in 2008 by
PALGRAVE MACMILLAN®
in the United States—a division of St. Martin's Press LLC,
175 Fifth Avenue, New York, NY 10010.

Where this book is distributed in the UK, Europe and the rest of the world,
this is by Palgrave Macmillan, a division of Macmillan Publishers Limited,
registered in England, company number 785998, of Houndmills,
Basingstoke, Hampshire RG21 6XS.

Palgrave Macmillan is the global academic imprint of the above companies
and has companies and representatives throughout the world.

Palgrave® and Macmillan® are registered trademarks in the United
States, the United Kingdom, Europe and other countries.

ISBN 978-1-349-37530-1   ISBN 978-0-230-61651-6 (eBook)
DOI 10.1057/9780230616516

Library of Congress Cataloging-in-Publication Data

Weaver-Hightower, Marcus B.
   The politics of policy in boys' education : getting boys right / by
  Marcus B. Weaver-Hightower.
     p. cm.
    Includes bibliographical references and index.

     1. Boys—Education—Australia. 2. Education and state—Australia.
  3. Boys—Education—United States. I. Title.

LC1398.A87W43 2008
371.823—dc22   2008015968

A catalogue record of the book is available from the British Library.

Design by Newgen Imaging Systems (P) Ltd., Chennai, India.

First edition: November 2008

10 9 8 7 6 5 4 3 2 1

The Politics of Policy in Boys' Education

Getting Boys "Right"

Marcus B. Weaver-Hightower

THE POLITICS OF POLICY IN BOYS' EDUCATION
Copyright © Marcus B. Weaver-Hightower, 2008.
Softcover reprint of the hardcover 1st edition 2008 978-0-230-60839-9
All rights reserved.

First published in 2008 by
PALGRAVE MACMILLAN®
in the United States—a division of St. Martin's Press LLC,
175 Fifth Avenue, New York, NY 10010.

Where this book is distributed in the UK, Europe and the rest of the world, this is by Palgrave Macmillan, a division of Macmillan Publishers Limited, registered in England, company number 785998, of Houndmills, Basingstoke, Hampshire RG21 6XS.

Palgrave Macmillan is the global academic imprint of the above companies and has companies and representatives throughout the world.

Palgrave® and Macmillan® are registered trademarks in the United States, the United Kingdom, Europe and other countries.

ISBN 978-1-349-37530-1   ISBN 978-0-230-61651-6 (eBook)
DOI 10.1057/9780230616516

Library of Congress Cataloging-in-Publication Data

Weaver-Hightower, Marcus B.
   The politics of policy in boys' education : getting boys right / by Marcus B. Weaver-Hightower.
     p. cm.
   Includes bibliographical references and index.

     1. Boys—Education—Australia. 2. Education and state—Australia. 3. Boys—Education—United States. I. Title.

LC1398.A87W43 2008
371.823—dc22   2008015968

A catalogue record of the book is available from the British Library.

Design by Newgen Imaging Systems (P) Ltd., Chennai, India.

First edition: November 2008

10 9 8 7 6 5 4 3 2 1

*To the memory of Matilda Faye,
beloved daughter.*

Contents

Illustrations ix
Foreword by Michael W. Apple xi
Acknowledgments xv
Abbreviations and Acronyms xvii
Notes on Formatting xix

Introduction 1

One   Gender and Education in a "New" Century: The "Boy Turn"   17

Two   Masculinity "Down Under": The Roots of Boys' Education Policy in Australia   29

Three   *Boys: Getting It Right*: Inventing Boys through Policy   59

Four   Means to an End: The Resulting Initiatives   103

Five   "Getting It Right" in the Schoolhouse: Two Case Studies   145

Six   Boys' Education in the United States: What Australia's Example Tells Us   179

Coda: Hope and Strategy   197

Appendix: Timeline of Australian Boys' Education Policy, 2000–2005   207
Notes   211
Bibliography   227
Index   243

Illustrations

Figures

I.1	The Cover of *Boys: Getting It Right*	13
2.1	The Political Map of Australia, Including Locations of Capital Cities and the 2000 Population Distribution	33
2.2	Timeline of Australian Gender Equity Policies, 1973–2002	46

Tables

I.1	List of Recommendations of *BGIR* by Acceptance or Rejection in the Government Response	7
3.1	Reduced List of Arguments from *BGIR* Used in Submissions Analysis	71
3.2	Cross-tabulation of Methods of Analysis for Agreement Between Informants and the Committee	73
4.1	Phase Two Boys' Education Lighthouse Schools, by State, Educational Sector, House of Representatives Electorate, Member of Parliament's (MP) Party, and Seat Status	130
4.2	Number of BELS Phase Two Schools, by State, Percent of Total BELS Schools, and 2001 State Percent of National Population	132
4.3	Declared Commonwealth Government Funding of Initiatives on Boys' Education, 2000–2005	137

Foreword

In *Educating the "Right" Way*, I devote much of my attention to the creative ideological work that the Right has done to shift the meaning of concepts that serve as a grounding for our economic, political, and cultural lives.[1] Words such as democracy, citizenship, equality, and similar things are what Raymond Williams rightly called "key words."[2] These are terms that have an emotional economy. They call forth a set of responses that tie us to or cause us to reject the policies and practices that dominate our societies.

Yet these words are also sliding signifiers. They can mean different things; and there is a constant struggle to fix their meanings so that they cohere with particular ideological assumptions and movements. Take the concept of "democracy." It can be either "thick" or "thin." It can refer to full common political participation in building and changing our institutions (thick) or refer to a more eviscerated sense of possessive individualism in which democracy is simply "consumer choice" on a market (thin). We can find an example of the thick version of democracy in education in the powerful reforms going on in Porto Alegre, Brazil, with their models of participatory budgeting and the Citizen School. The thin understanding is evident in proposals for voucher plans and marketized models of schooling, with their vision of consumer choice as the guarantor of "democracy."[3]

Marcus Weaver-Hightower takes one of these key words as the backdrop of his fine analysis in this book—the word "equality." He examines a powerful issue that has been taken up largely by rightist forces, although it doesn't have to be only these groups who are activists in dealing with the problem. The issue is boys and their education. For years, supposedly, educators focused on making schooling and its practices and results more equal for girls and women. Increasingly, however, a question is raised, often at a fever pitch: What about the boys? How can we have equality when boys and the problems associated with them are ignored?

The focus of this book is on Australia, the first nation of its type to actually have a federal policy on the education of boys. But the centrality of the question of "What about the boys?" is clear in the United States and other nations as well. Thus, Christine Hoff Sommers' popular book *The War Against Boys* is just one of many recent volumes that have lamented the sad state of boys and the utter neglect of their problems in our educational system. Weaver-Hightower takes his cogent analysis of Australia and applies it to the United States, demonstrating that Sommers and similar authors get it wrong, not right.

There is no doubt that there are problems and many more progressive authors have examined what it is like, for instance, to be a black young man in all too many schools here.[4] As the father of an African American son myself, I can give ample personal testimony to the dangers—and immense structured racism—constantly faced by black male youth.

As a clearly profeminist man, one of the strengths of Weaver-Hightower's account is that he does *not* assume that "women's issues" have been "solved." This is not a zero-sum game for him. Nor does he buy into the rightists' articulations of these issues. He correctly sees that the issue also concerns particular forms of masculinities. In so doing, he builds upon some of the fine work that has been done by Connell, Mills, Lingard, Lingard and Douglas, and others.[5] All of them understand that gender is a relational concept. Thinking about issues surrounding women's ongoing struggles requires that we also think about sexuality and the body, about compulsory hetero-normativity, about forms of masculinities, and about the ways in which all of these are organized in and around dominant institutions.[6] And since, as we know, the state itself is fully gendered, this gives Weaver-Hightower's focus on state education policy and how it reproduces and interrupts dominant gender codes even more importance.

The existence of "boy panic" is not new in many of our nations, of course. Indeed, speaking honestly and personally, my own beginnings as a teacher in the slums of a declining industrial city on the east coast of the United States have some connection to this phenomenon. In the early 1960s, I had just gotten out of the army and had completed less than two years of courses at a small teachers college. But the army had "trained" me as a teacher (and truck driver—don't ask how these things got sutured together in their minds). Yet, even without a degree, I was given an emergency teacher certification and was hired as a full-time substitute to teach in inner city schools that had a reputation as being truly tough places. Getting more men in these classrooms was then (and now) seen as a solution to the problem of undisciplined working class and poor students—mostly students of color. Black and brown boys were seen as problems. I was a man; I had been in the army. People like me were "the answer." And at the not very generous pay of $15 a day with no benefits, I was a cheap answer.

Just as "boy panic" is not only a recent phenomenon, critical work on boys is not new to critical cultural research in education either. Some of the most germinal work in the history of critical educational studies that has been interpreted as being about class relations has actually also been about the ways in which masculinities are produced in and around schools.[7] This is also the tradition in which this book stands. But how it does this advances these kinds of analyses in important directions.

In an early piece on the state of cultural studies, Richard Johnson lays out a framework for doing substantive cultural analyses.[8] He suggests that there is a *circuit of cultural production* that has three "moments"—production, distribution, and reception. To fully understand the politics and possibilities of any product, policy, or practice one would have to situate it into the conditions that produced it, how it was made available, and how it was received. None of these moments can be easily reduced to the others. And each moment can be the site for political intervention.[9]

In many ways, with its focus on what the author calls "policy ecology," *The Politics of Policy in Boys' Education* is grounded in a recognition of the circuit of cultural production. It engages in a detailed and nuanced critical analysis of the social and ideological conditions for the development of a national policy on boy's education, shows how it was made available and funded, and illuminates what happened when it met the realities of actual schools, teachers, and students. A critical examination of how these various moments relate, mediate, transform, or even reject a policy is more than a little unusual in the literature on policy formation within the critical educational studies community.[10]

The Politics of Policy in Boys' Education also recognizes that some of the concerns about the education of boys have been progressive, not only retrogressive. As Weaver-Hightower argues:

> Boys' education issues are neither inherently conservative nor inherently dangerous. Indeed, working with boys has tremendous potential for progressive, socially just education. Thus far, however, rightist movements and conservative authors have indeed co-opted these issues. Their conservative adherents don't, though, permanently taint such issues. With care and alertness, interventions can serve the interests of both boys and the society they help make up.

This quotation speaks to one of the significant aspects of Weaver-Hightower's efforts here. He isn't content to understand the social and ideological context of the policy, how it was pushed (or not) by the Government, and what happened at the level of real schools—although this in itself is a real accomplishment. He also wants to go further, to assess what we in Australia, the United States, and elsewhere can learn from this. Just as importantly, Weaver-Hightower lays out important strategic suggestions for activists in the profeminist communities both to interrupt the rightist movements that have occupied the space of this issue and to go further in organizing educational policies and practices that take the issue in more progressive directions.

In essence, what he has done is to take seriously a number of what I have called the tasks of the scholar activist in education.[11] Among these tasks are: bearing witness to negativity; showing spaces for possible counter-hegemonic work; keeping alive multiple critical traditions and building new ones on the basis of multiple dynamics of power; and acting in concert with progressive forces that seek to interrupt dominance. *The Politics of Policy in Boys' Education* succeeds in engaging seriously with each of these tasks, thereby giving us even more reasons to read it and learn from it.

MICHAEL W. APPLE
John Bascom Professor of Curriculum and
Instruction and Educational Policy Studies
University of Wisconsin, Madison

Acknowledgments

This book is the product of help and sacrifice by numerous people. In little or no particular order, I thank: my parents, Faye and O'Neal Hightower; my brother, Anthony; my grandparents, great grandparents, aunts, and uncles; the traditional owners of the lands on which my education and research took place, in both the United States and Australia; taxpayers in Wisconsin, North Dakota, the United States and Australia who in various ways indirectly funded my education and research; the Australian-American Fulbright Commission, including Mark Darby, Judith Gamble, and Sandra Lambert; Kevin Jackman, President of the Queensland Fulbright Alumni Association; my doctoral committee members, Amy Stambach, Deborah Brandt, Francois Tochon, and Diana Hess; Bob Lingard and Martin Mills for friendship and logistical support; the Australian House of Representatives Education and Training Committee Secretariat, particularly Gaye Milner and James Rees; Jack Weaver and Laura Garner; David Bloome; Deborah Rowe; my colleagues in Friday Group, especially my dear friend Bekisizwe Ndimande; my current and former students from Goose Creek High School, Vanderbilt, and the University of North Dakota; and colleagues including Fazal Rizvi, Julie McCleod, Jeff Sun, Clint Hosford, Jason Lane, Eric Wolfe, and Kathleen Gershman (who sacrificed her time to protect mine).

Special mention should be made of several people. Michael Apple, my friend and doctoral advisor, through his willingness to suspend disbelief, his trust in me to help with his own projects, and his friendship during my education, has been a tremendous influence on my life. His contributions are among the greatest to the following pages. Matilda, my "angel baby" to whom I dedicate this book, has been my teacher of what is truly important in the universe. Harrison, my living baby, is a cuddly reminder of the implications of governments making policies for boys; I want his to be a life lived completely and humanely. More than anyone, though, I owe thanks to my wife, Rebecca. As amazing as her love for me is, her willingness to follow me (literally) around the world, to delay her own best interests, to read and comment on "lists of names," and to generally put up with me are even more spectacular. She is the greatest combination of scholar, person, and friend that I have had the pleasure to encounter.

* * *

Portions of this book have appeared in different forms elsewhere. A portion of the Introduction is derived from my article "An Ecology Metaphor for Educational Policy Analysis: A Call to Complexity" in *Educational Researcher* (Volume 37, number 3). Chapter 1, on the boy turn in education research, is derived from my article "The 'Boy Turn' in Research on Gender and Education" in *Review of Educational Research* (volume 73, number 4). Chapter 6, the comparative view of the United States, derives from my chapter in Wayne Martino, Michael Kehler, and my volume, *The Problem with Boys: Beyond the Backlash in Boys' Education* (Binghamton, NY: Routledge, forthcoming).

Abbreviations and Acronyms

ACSSO	Australian Council of State School Organizations
AEU	Australian Education Union
ALP	Australian Labor Party
AAUW	American Association of University Women
B2FM	Boys to Fine Men conference
BELS	Boys' Education Lighthouse Schools
BGIR	*Boys: Getting it Right*
DEST	Department of Education, Science and Training
DETYA	Department of Education, Training and Youth Affairs
EDT	House of Representatives Standing Committee on Education and Training
EQ	Education Queensland
GEF	*Gender Equity Framework* [refers to *Gender Equity: A Framework for Australian Schools*]
GSI	GaiSheridan International
HECS	Higher Education Contribution Scheme
HREOC	Human Rights and Equal Opportunity Commission
ICTs	Information and Communication Technologies
MCEETYA	Ministerial Council for Education, Employment, Training, and Youth Affairs
MP	Member of Parliament
MULTILIT	Making Up Lost Time In Literacy
NAEP	National Assessment of Educational Progress
NCLB	No Child Left Behind Act of 2002
NSW	New South Wales
NT	Northern Territory
NTDEET	Northern Territory Department of Employment, Education & Training
OP	Overall Placement rank
PSD	Personal and Spiritual Development

RSC	"Riverside Schools Cooperative"
SDA	Sex Discrimination Act
SA	South Australia
SFB	Success for Boys Program
SMT	Student Management Team
SRS	"Springtown Religious School"

Notes on Formatting

All quoted material retains the original spelling and punctuation. References to the "*Hansard*," the transcripts of Australian Parliamentary hearings, are specifically to the *Hansards* of the House of Representatives' Inquiry into the Education of Boys hearings. References to specific pages use the running page count listed in the Hansard; for entire Hansards from a particular hearing, I have referred to them by date. In referencing the Australian House of Representatives' Committee on Education and Training's Inquiry into the Education of Boys, I have opted to refer to just the "Committee" and the "Inquiry"—capitalized—for the sake of economy. Finally, Australian currencies (the dollar, given as "A$") have been given the U.S. equivalent ("US$") in parentheses for comparison. I have used the late 2007 conversion rate of A$1=US$0.872, though this rate is different from the time of the research (the Australian currency has improved about 15 cents against the U.S. dollar since 2003). I have also rounded to the nearest dollar.

Introduction

> *Give Australians an issue and they will argue it so passionately and in such detail, from so many angles, with the introduction of so many loosely connected side issues, that it soon becomes impenetrable to the outsider.*
> —Bill Bryson, *Down Under*, 343

Australia can be a land of extremes. Drive westward from the subtropical hills of Brisbane, covered in lush foliage and year-round blooming trees, and in a couple of hours you will be in sparse, semiarid dessert. Drive a day in that same direction without water, and abandon all hope, for you will be right in the middle of the "Red Centre," with vast surrounding miles of little more than rock and dust.

Extreme, too, are the creatures that crawl upon this continent. Though famous for its cuddly koalas (though you wouldn't say "cuddly" if you ran afoul of one in the wild, and *don't* call them "bears"), Australia also has more animals that will kill you—more deadly snakes and spiders—than any other place on Earth. Australia even hosts a "stinging" *tree* that can kill you if you stumble into its branches. Even with all those things that can snuff out your life, there are no big predators. The dingo, not to be confused with a dog, though it looks much the same (and they *probably* do eat babies), is the largest nonhuman predator on the continent. No lions, no tigers, and no bears.

The people can bear remarkable extremes, too. Many people note Australians as one of the most generous and friendly people on Earth. I count myself among those. At the other end of the spectrum, though, Australia harbors a dark racial history, including genocide, massacres, the stealing of land, stealing children from parents, assimilation schools, and legislated prejudice that denied Aboriginals and immigrants suffrage and even the right to full wages. Even as I write this, Australia has one of the most restrictive immigration policies of a "first world" nation. Until 2003, in fact, scores of children were put behind high fences in a desert detention camp in Woomera, South Australia, where abuse was reportedly rampant. Moreover, in a land that prides itself on "the fair go"—a kind of meritocracy myth—Australia's indigenous populations have some of the poorest health, education, and unemployment indicators of any settler society. Clearly, Australia isn't alone in such a racial history—nearly all these same things can be said of my home country, the United States' racial history—but the contrast seems to stand in higher relief here amongst the jocular larrikin blokes who call you "mate" the first time they meet you, at least if you're a white man.

In a land of extremes, it seems only natural that gender equity should be a polarizing debate. Currently, the debate is all the more polarizing because the gender equity that so many Australians are concerned about is "equity" for boys.

Beginning in the early 1990s, a few Australian educators and politicians became concerned about the educational and social indicators of Australian boys. Two decades of pioneering work on educational policy for girls had raised awareness of gender issues, and advocates increasingly began asking, "What about the boys?" What about boys' literacy scores, which—like girls' math scores—showed boys lagging? What about the greater number of boys dropping out of school? What about boys' higher rates of risk taking, car crashes, drug use, and suicide? What about the lack of male role models and the "feminized" environment in schools? Such questions combine a mix of genuine concern and a backlash politics against the education gains of feminism.

Answers to such questions were quick and many from feminist scholars and educators. Girls were still not achieving parity with boys in math, science, and technology. Girls and women, despite literacy advantages, were not as able to convert the advantage into job market or political success. Violence and sexism against women were still rife. Importantly, also, to the question "What about the boys?" came the Socratic answer, "Which boys?" That is, a counter-discourse developed: rather than *all* boys being at risk, we were reminded that certain boys—Aboriginal boys, working-class boys, nonheterosexual boys—are disproportionately affected by the problems identified by boy advocates.

These "sides" as I have described them are admittedly cartoonish and without the nuance of the actual debates. These are the kernel positions, though, and for years these have remained intractable, extreme. As the epigraph above from Bill Bryson suggests, the many positions have been argued passionately and in great detail, the angles from which they are discussed have metastasized, and numerous side issues—from teacher training to male-only scholarships to phonics and so on—have been raised. Where I disagree with Bryson is that such arguments are not impenetrable to the outsider. That is where this book comes in.

This book details a unique occurrence on a continent that has become synonymous with uniqueness. Rather than flora, fauna, or sports in this case, though, I focus on a unique event in the worldwide boys' education debates: the development of a *national* boys' education policy. The debates themselves are not a solely Australian phenomenon; countries worldwide are struggling with these issues, which I detail in the first chapter and in comparing Australia with the United States in chapter 6. The extent of the debates and the formation of national-level policy, however, is uniquely Australian.

Given that boys' underachievement has been evident in Australia and other countries for a long time, why are public panics occurring now? I suggest that, first, boys' education debates are not new, not even to Australia, and, second, that boy panics like the current one happen as a response to specific and often identifiable cultural anxieties. Furthermore, such panics are cyclical, reoccurring at intervals as masculinity comes under fire or drastic social upheavals occur. This is evident, for example, in the United States, where late nineteenth century "boy problems" were also fretted over, where the socially tumultuous 1960s saw panics over female teachers making boys too feminine, and where present panics over boys follow moderately successful reform for girls' education.[1] This book is an attempt to understand the genesis and anxieties behind the most recent worldwide boy panic.

Though other nations have engaged boy debates, in this newest "wave" dating from the early 1990s, Australia stands apart in the quantity of discussion, its visibility, and the lengths to which it has gone. Particularly unique about Australia, more than just public awareness, is that boy debates there have caught the attention of national policymakers. The Australian House of Representatives released a report, *Boys: Getting It Right* (hereafter *BGIR*), a document that has led to federal policy on boys' education.[2] *No other country has yet formulated federal-level policy on the education of boys.* This makes Australia a first and—at this writing—the only one. It isn't, though, an outlier with nothing to teach other countries.

But why study Australia? Why should scholars in the United States, England, Japan, or Iceland care about what happens thousands of miles away in a country of only about 20 million people? While all may learn something different from the Australian experiment in boy policy, there are important reasons to examine this unique occurrence. First, Australia's example can indicate, through its similarities and differences, what may happen elsewhere when boy-panics and -debates occur—the "Could this happen here?" question. Why has Australia gone so far with boys' education issues and other countries have merely metaphorically wrung their hands? In the final chapter, I discuss the United States as an example of analyzing the "fit" of Australia's experience, explaining why a concerted, codified policy formation has not occurred in the United States—yet—and why such policy may be difficult to create. Still, other aspects of the Australian example—say, the gendered shifting of research funding and the impact of federalism—are highly relevant to the United States and serve as warnings as researchers, policymakers, and educators proceed.

A second reason to study Australia's boys' education policy is that educational discourses are highly globalized. The "flows" of educational ideologies are truly complex, and they are inextricably tied to the political alliances that traverse national boundaries.[3] Just as the conservative Howard Government in Australia could borrow flying the national flag at schools as a wedge issue from U.S. politicians, there is little to prevent U.S. politicians from borrowing boys' education as an issue for scoring political points. Other countries' debates over boys' education have *already* influenced and fed media and bestseller panics in the United States, and are being used as rationale for reforming boys' education.[4] What has happened in Australia, then, becomes a key learning tool to those who must respond should advocates in the United States pursue the issue.

Finally, I believe educators need a deep understanding of cross-cultural education if they are to find the best way to educate students.[5] No country can claim that its education system has achieved perfection in all aspects, but some do better than others in identifiable areas. Studying Australia, or any country other than one's own, shows how things might be different, why shared similarities work or don't work, and different methods to educate in radically more progressive ways. Overlooking international systems—whether from ignorance or arrogance—imperils the ability to equitably educate future generations.

Still, despite the benefits of comparisons, a caveat needs to be made. This book does not *primarily* seek to be a comparative study. Rather, this book, as an empirical study, focuses mainly on Australia. I have included an explicitly comparative chapter at the end to further elucidate the importance of the novel policy Australia has undertaken. I have also included short sections at the ends of the Australia-focused chapters that suggest applications to the United States and other contexts. Readers are encouraged to find in that comparative chapter and the applications sections a method for assessing Australia's example for their own context. More than this comparative use, though, I hope readers find in the following pages useful ways of viewing policy, masculinity politics, and gendered education.

Boys: Getting It Right and the Government Response

This book centers on understanding the origins, meaning, and implications of a single policy, *BGIR*. It might be helpful to readers in approaching the rest of this book, then, to have a basic understanding of what *BGIR* actually says. The full report is available online, and I return to a deeper analysis throughout, but following is a briefing on *BGIR* and the government response to it.[6]

The Report, *Boys: Getting It Right*

BGIR consists of seven chapters and numerous pieces of front and back matter that describe the report and the Inquiry process. The report's first chapter is an introduction that, like the executive summary before it, generally articulates the reasons for taking up the Inquiry—parental and teacher concern as well as costs to society and boys themselves of continued "under-achievement"—and the general arguments of each upcoming chapter. Chapter 2, "Stating the Case: School and Post-school Outcomes," gives a range of measures along which the Committee asserts that boys are disadvantaged, including early literacy attainment; numbers staying on until Year 12 and dropouts, suspensions, and truancy; Year 12 performance and gendered subject choices; and social outcomes, employment, and higher education enrollment. The chapter also includes a lengthy section on indicators for Aboriginal and Torres Strait Islander students. In general, the indicators in the chapter are employed in asserting *dis*advantages for boys, though some indicators they use—particularly employment indicators—suggest that boys do as well or better than girls in some respects.

Chapter 3 provides reasons and causes the Committee asserts for the "decline" and underachievement of boys. These include (a) labor market changes, like the increase in part-time versus full-time work and the disappearance of many traditionally male-oriented trade jobs; (b) social changes, such as the increased status of women, more single parent families and absent fathers, and increased negative media images of males and masculinity; and (c) gender equity policy that has focused solely on girls or that treat boys' needs only insofar as they must change to help girls.

Chapters 4, 5, and 6 of *BGIR* interrogate what schools have done to disadvantage boys, and they present a vision of schooling that the Committee asserts would be more beneficial to boys. Chapter 4 addresses curriculum, pedagogy, and assessment, claiming that boys suffer from an irrelevant, boring, and inappropriate curriculum to their supposedly innate learning style (77–79). It also suggests numerous methods and strategies that "work in practice," including common assertions about boys' love of competition; boys' particular need for rules, structure and physical activity; and the integral role of praise—all reactionary arguments against the perceived "forgetting" of classic, time-tested teaching methods.

BGIR's Chapter 5 focuses on literacy and numeracy (an Australianism for mathematical literacy), though they emphasize boys' traditionally weaker area of literacy. The Committee first explores explanations for boys' lower literacy performance, like behavioral difficulties and developmental lags compared to girls, the lack of phonics instruction, and greater incidences of hearing and auditory processing difficulties. The Committee makes numerous suggestions for fixes, including providing a literacy coordinator, installing phonics into all in-service and preservice teacher education, and reducing class sizes.

Chapter 6 then explores the contentious but often mentioned argument for having more male role models, particularly male teachers. The Committee suggests that peers be used to help boys in school. They also assert an ineffable uniqueness to fathers' and male teachers' contributions, particularly for boys. Finally, the chapter provides an exploration of various school structures, particularly middle schools (a relatively new and increasingly popular arrangement in Australia), alternative schools, and single-sex schooling.

Chapter 7 of *BGIR,* the shortest at three pages, is perhaps the most obviously political, invoking federalist tensions between the Commonwealth and the state and territory governments, which I detail in my chapter 2. All three recommendations from the chapter (recommendations 22–24) seek to further define the Commonwealth as supervisor of actions taken by the states and territories, so their implications go beyond boys' education. Recommendation 22 seeks to compel states and territories to provide specific information about their schooling, lest the Commonwealth turn the job over to the Australian Bureau of Statistics. Recommendation 23 seeks to accomplish what was a favorite policy initiative of the Commonwealth under Minister Nelson: making the state systems more consistent and comparable, more uniform. Recommendation 24 sets the stage for the Commonwealth escaping blame if policy initiatives go underfunded by suggesting *in advance* that states might try to reduce their own contributions if the Commonwealth gives them more money.

The Government Response

In Australia, a report only becomes policy (a notion I trouble below) when the Government minister in that policy area responds to the report, giving a statement of what actions the Government (capitalized in the Australian context when referring to the political party in charge; the other party is "the Opposition") intends to take. For the boys' education inquiry, the Government response is *Boys' Education: Building on Successful Practice* (hereafter just "the Government response").[7] Its reaction was generally positive to *BGIR*, for the Government supported most recommendations made and the intent of several others (see Table I.1). Even some recommendations rejected in the Government's response were later pursued through other policy means. These include the bill to amend the Sex Discrimination Act (SDA) to implement male teacher scholarships (recommendation 20; see my chapter 4) and the National Inquiry into the Teaching of Literacy to retry requiring phonics (recommendations 7, 8, and 9). In keeping with the legal nature of the document, the explanations that are given for rejection or acceptance of each of the 24 recommendations involve describing the limitations of funding or the conflicts of policy or law that prevent acceptance of the few suggested initiatives that the Government rejected. In the end, the only truly rejected recommendations were the more progressive recommendations for better labor conditions for teachers (raising pay [recommendation 18] and lowering class sizes [13]) and health services for students (hearing and vision screenings [5]). Others rejected include compelling certain statistical reporting (recommendation 22) and measures to prevent states and territories from lowering their financial contributions (recommendation 24).

Table I.1 List of Recommendations of *BGIR* by Acceptance or Rejection in the Government Response

No.	[Recommendation]
Accepted Recommendations	
1.	Recast the national *gender equity framework*
2.	Preservice and inservice training on learning styles and needs of boys and girls, with a focus on pedagogies that help boys
3.	More research on impact of different assessment methods
4.	Awareness campaign on impact of parenting styles
6.	Public and childcare awareness campaigns on needs for pre-literacy and pre-numeracy skill development
11.	Commonwealth should ensure Literacy and Numeracy program funding is being used effectively
14.	More research in engagement and motivation

Continued

Table I.1 Continued

No.	Recommendation
15.	Fund comparative research on different school structures, curricula, assessments, and senior school alternatives
16.	Fund the assessment of state and territory programs for the disadvantaged
17.	Encourage teacher education to focus on behavior management and interpersonal skills; provide same as professional development for inservice teachers
21.	Promote strategies for involving fathers and men
22.	Review existing educational data; consider making this review a regular Australian Bureau of Statistics role
23.	MCEETYA seek policy and assessment consistency and comparability between states and territories
24.	Monitor Commonwealth-funded programs and ensure states don't reduce their contributions

Rejected Recommendations

No.	Recommendation
5.	Provide hearing and vision screening and implement the Victorian auditory processing program
7.	Commonwealth-funded literacy programs should include intensive, explicit phonics
8.	Teacher education should increase focus on literacy and numeracy pedagogy and include training in phonics
9.	Funding for professional development should be increased to target literacy and at-risk students and should include phonics training
10.	Provide a literacy coordinator and early intervention literacy specialist at every primary school
12.	Provide preservice and inservice training for secondary teachers in literacy issues
13.	Reduce class sizes
18.	Increase teacher pay, especially for experienced teachers
19.	Expand requirements for teacher training admission that includes more than academic success
20.	HECS-free scholarships for equal number of males and female

Conceptualizing Policy

How analysts conceptualize policy and policy processes shapes the interpretations they make. I view policy as a highly complex, often contradictory process that defies the cleanliness and singular purpose that the popular imagination holds. The traditional view asserts that the policy process functions rationally, usually following a problem→research→solution→implementation model.[8] This rational model, often called the "stages heuristic," mostly developed in the 1960s to assist governments with supposedly technically sound policy formulation and resource allocation.[9] It holds that solving problems requires finding the one likely solution and using that policy lever to make predictable and efficient changes. Such a view grossly misjudges the complexity and grittiness, the false starts, unabashed greed, and crashing failures of some policy formation and implementation.

Analyses of the policy process, particularly since the 1980s, have increasingly moved away from such views toward more complex, post-structural, and postmodern perspectives. Stephen Ball perhaps best summarizes this viewpoint:

> National policy making is inevitably a process of *bricolage*: a matter of borrowing and copying bits and pieces of ideas from elsewhere, drawing upon and amending locally tried and tested approaches, cannibalising theories, research, trends and fashions and not infrequently flailing around for anything at all that looks as though it might work. Most policies are ramshackle, compromise, hit and miss affairs, that are reworked, tinkered with, nuanced and inflected through complex processes of influence, text production, dissemination and, ultimately, re-creation in contexts of practice...[10]

Others have rightly argued that traditional views of policy as neutral and in the best democratic interest belie the true impact of policies. Prunty, for example, in defining critical policy analysis, ruptures the notion that policy can be value-free and that its purpose is simply the smooth functioning of the state and its institutions.[11] Policy, in his view, serves the interests of specific people, usually the already powerful.

In this book, I forward a qualitative view of policy that builds on these post-structural, postmodern, and critical insights. I conceptualize policy in what I have elsewhere described as a policy ecology metaphor, a metaphor that encapsulates the messy workings of power relations along with the inflections of history, culture, economy, and social change.[12]

A *policy* ecology acts in similar ways as biological ecosystems. A policy ecology represents the combination of a policy (or a related group of policies) with the texts, histories, people, places, groups, traditions, economic and political conditions, institutions, and relationships that affect it or that it affects. Every contextual factor and person contributing to or influenced by a policy, both before and after and in any capacity, exists within a complex ecology. This demands analysis outside of just the politicians who construct a policy or the teachers who enact it (or don't); an ecology demands examining media, parent groups, printers, travel agents, spouses, and all other persons or institutions that allow the process to work. It also necessitates understanding the broader cultures and society in which the policy resides. Limits, of course, are necessary on how deep any analysis can go, but the ecology metaphor's usefulness lies in its ability to extend analysis further. Using this policy ecology conceptualization, then, the remainder of *The Politics of Policy in Boys' Education* explores the many important actors, relationships, environments, and processes of the boy debates in Australia before, during, and after the construction of *BGIR*.

Of course, treating *BGIR* as—and calling it—a "policy" puts me on contested terrain. Indeed, during interviews if I referred to it as a "policy," several interviewees quickly and explicitly corrected me. For example, I asked one of the MPs on the Committee, Sid Sidebottom, how he thinks they will gauge whether "the policy" has been effective, and he responded, "Well it's not a policy actually. Right. Gosh, by the time this policy's made I'll probably be run out of town." Sidebottom here refers to the technical distinction that analysts and policymakers in Australia make between a "policy" and an inquiry "report." In a typical inquiry (described in chapter 3), the Committee prepares a report on the findings of its inquiry for the Minister for Education, Science and Training, and then the bureaucracy of the Ministry—the Department of Education, Science and Training—formulates the official policy in response to the report. The Government response to *BGIR*, described above, technically speaking, represents the "official" government position, the "policy."

I argue that, despite the technically incorrect nomenclature, *BGIR* does in fact represent a "policy," not just a "report" (a word that underestimates the true impact of *BGIR*). True, *BGIR* does not have an enforceable mandate, nor does it have the authority in and of itself to direct governmental funding; only the minister and his or her response to the report can do that. *BGIR*, however, in many ways functions *as if* it were policy. Several concepts support this argument.

First, Stephen Ball lays out a conceptualization of policy as dual: as *both* a text and a discourse.[13] As text, the policy exists as a concrete, analyzable document that can be read. The traditional analyst, one from the realist or rationalist school, can easily operate at this level of analysis. Policy as discourse, conversely, entails how policies "exercise power through a *production* of 'truth' and 'knowledge'" (21). In other words, policies bid to control the "thinkable," the "facts," and who counts as "expert." With the imprimatur and legitimacy of the state behind it, not to mention the state's massive funding, government policy wields significant control over what can be thought and said.

Related to this, Lingard argues that some texts that are not, in fact, "policies" per se—in his case, popular press books on boys—can become "de facto policy" in a context of weak state control over education policy and increasing site-based management.[14] In other words, practitioners, when formal state policy does not exist, look to other sources to solve their daily problems. They look to popular books or reports to fill the gaps in policy, thus allowing those texts to become a kind of policy.

Both observations, of policy as discourse and of de facto policies in the absence of formal guidelines, apply well to *BGIR* and make it appropriate to call it, quite accurately, a "policy." Taking the latter first, while Australian policy on boys' education has existed to a small degree at the state level in New South Wales (see chapter 2), the other states and territories have had little formal policy, and no policy on boys specifically existed at the Commonwealth (that is, national) level until this Inquiry. This has allowed (or necessitated) many schools to take up this report as de facto policy. The case studies in chapter 5 are prime examples.

To the policy as discourse, while not uncritically adopted word for word—for no implementation happens in a straightforward way—elements of the report have been taken as "official knowledge" by certain schools.[15] They take the report's endorsement of, say, the Spalding method—a phonics-intensive approach—as the way reading should be taught, for who, some might wonder, can find experts if a government inquiry cannot? By their mere citation in the report, too, some scholars gain expert status. These scholars reap the benefits of this expert status once the dollars begin flowing from the Commonwealth. This is the essence of policy as discourse. The report establishes what counts as "thinkable" and those authorized to speak it.

Still, this raises the question: When does a "report" become a "policy"?[16] Is some vague notion of influence the major determiner, and, if so, how much influence must a "report" have to finally be considered a "policy"? Take an example from the United States when a "report" might clearly be thought of as a "policy": the report *A Nation at Risk*.[17] Perhaps no other piece of writing has more profoundly affected U.S. education. This single report helped to establish the dominance of economic competitiveness as a reason for educational reform, the perception of U.S. schools as failing, and the need for standards and accountability as *the* fix for education's problems. This report led to actual policy in state after state despite relatively weak federal control of education in the United States, and the report shapes the Republican Party's (and to some extent the Democrats') education platform even today. *A Nation at Risk,* though only technically a "report," has such major influence that one would certainly be justified to call it a "policy" that has been implemented in education for more than 20 years.

While *BGIR* certainly does not enjoy the same amount of influence, its impact has been substantial. On the international level, it has been cited in the United Nations Children's Fund's report, *The State of the World's Children 2004*, as a major government response to gender inequalities.[18] In Australia the report (*not* the Government response, which is technically the "policy") has been cited in official arguments over changing sex discrimination laws to allow for male-only teaching scholarships, arguments for Labor's policies for male mentoring schemes, arguments for implementing single-sex education, and arguments for conducting a national inquiry into reading.[19] Perhaps most directly, as I show in chapter 4, it has even provided the guiding mission for rewriting the national *Gender Equity Framework* (*GEF*), the nation's longstanding gender and education policy. *BGIR* directed not only *that* it be done, but also *how* it would be done and with what ends. Again, the Government response—what technically is the "policy"—rarely receives a mention. Instead, *BGIR* is cited and is acknowledged as legitimate, official, and actionable information. It has become discourse.

BGIR has also become de facto policy, for local institutions have looked to it for advice and information in the absence of other policy. Numerous schools cited it in newsletters, some even using its findings as advertising for their own schools or programs.[20] Springtown Religious School, one of the case study schools in chapter 5, publicized *BGIR* in communication with parents; its Parent and Friends Association discussed the findings; and Mrs. Howard, one of the deputy principals, used it as a major source for her research on resilience.

As text, the form and layout of the document is reminiscent of a policy, aiding the slippage between "report" and "policy," a slippage that could increase its legitimacy among those who don't know that it is "just" a report. Like other actual policies, *BGIR* is nicely printed and typeset, it uses reference to government agencies and figures (the cover proclaims the imprimatur of "The Parliament of the Commonwealth of Australia" alongside the national seal; see figure I.1), it cites experts and authorities, it contains official statistical tables, and it even numbers paragraphs for easy citation, just as if it were a statute or law. It *looks like* a policy, for it looks like the things called "policy" that have come before it. That someone could misrecognize the report as policy—both in the folk sense of "mistake for" and in the Bourdieuian sense of intentional construction to hide the truth of the social structure[21]—isn't entirely unintended or unwelcome for the Committee's purposes of establishing their interpretation as dominant and correct.

Figure I.1 The Cover of *Boys: Getting It Right*

Source: The cover of the report is reproduced with the permission of the Australian Department of the House of Representatives.

14 / POLITICS OF POLICY IN BOYS' EDUCATION

All of these factors considered, *BGIR* acts in the capacity of policy at the levels of text, discourse, and de facto policy. Thus, while the *BGIR* report might not fulfill the government-established definition of policy, in both symbolic and real ways *it is a policy.* In the rest of this book, therefore, I regard *BGIR* as a policy, and I treat it to the same analyses that an "official" policy would receive.

Book Overview

To reduce this book to its simplest, again, this is the story of a single policy. It is also, though, much more than that. From its origins in the history of gender and education policy in Australia to the resulting initiatives that focus resources on the problems of boys, this book examines a vast array of events, people, arguments, and programs that revolve around the Australian House of Representatives Standing Committee on Education and Training's *Boys: Getting It Right,* with particular focus on the period 2000 to 2005. More than this, though, through this book, I theorize *BGIR's* importance and the larger discourses it speaks through as key moments of masculinity politics with ramifications worldwide.

Based on a yearlong qualitative study in Australia and two subsequent trips—including stakeholder interviews and participant-observation in schools and with politicians—and years of documentary analysis, *The Politics of Policy in Boys' Education* moves from the macro- toward the micro-level of the worldwide boys' education ecology.[22] I start at the widest level in chapter 1, giving an overview of the "boy turn" in gender and education research and policy internationally, a force that has driven Australia's boys' education debate and many like it worldwide. Moving one level closer, in chapter 2 I give the cultural, political, and policy environment for the Inquiry by detailing the context and history of Australia, including the previous work on gender and education there. Then, moving even closer in chapter 3, I examine the process of the Inquiry into the Education of Boys, its report *BGIR,* and the Government response. I then document and analyze the various programs and initiatives that have flowed from the Inquiry and its report in chapter 4. Finally, arriving "on the ground," where education policy "happens" in schools, chapter 5 presents two cases of real schools and educators dealing with (*not* just "implementing") the policy. Then, chapter 6 explores what Australia's example means comparatively by focusing on the United States' boy debates. I then end the book with a Coda that lays out possible strategies of resistance and reasons to hope for those wary and weary of boys' education issues. That is the basic outline, but, like Bryson's description in the epigraph of any good Australian argument, there will be much greater detail, many angles, and more than a few side issues to work through to truly understand the depth and breadth of this single policy.

CHAPTER ONE

GENDER AND EDUCATION IN A "NEW" CENTURY: THE "BOY TURN"

This chapter explores the international context for boys' education policy. *Boys: Getting It Right* (*BGIR*), the Australian policy at the center of my analysis, may have been the first policy of its kind, but it didn't come from nowhere. Rather, it is a product of worldwide concerns about boys' education that have enveloped, among others, Canada, England, Germany, Iceland, Japan, and the United States. Viewing it this way, this chapter describes what I call the "boy turn" in gender and education research and its etiologies, overviews boys' advocates' major concerns, and, finally, discusses the implications of the boy turn on the development of *BGIR*.

Turning to Boys

The landscape of gender and education has become quite complex. Where educationalists once focused, rightly, on the massive inequalities for girls in schools, the focus now has turned to boys. Because boys and masculinities have become such pervasive topics, addressed by hard-line boy advocates and feminist educationalists alike, one can best describe the current state of the debates as a "boy turn."[1] At present, scholars and educators in numerous countries are working hard to answer the frequent, often conservative or antifeminist question, "what about the boys?" and funding pours in globally in ways boy advocates of a decade ago could have only dreamed. In Australia alone, as I document later, millions of dollars have been granted to both theory- and practice-oriented boyswork.[2] Most programs for girls are not faring nearly so well.

Let us not overstate, though. No movement is totalizing or monolithic, facing no resistance. Girls' education debates never were, and even for all the patriarchal force behind it, the boy turn isn't. Still, for those on the bottom looking up, it surely seems daunting, this pervasive, ubiquitous concern for boys. How did concerns over boys get where they are, and what issues cause everyone so much concern? What has caused this "boy turn"?

What about the Boys?

While worries have been vast (boys are, after all, about half of the population), alarm over boys' schooling generally fall into two broad categories: (a) academic achievement and learning and (b) social, personal, and economic outcomes. The academic achievement and learning category is perhaps the narrowest, but it tends to grab many headlines. This partly stems from recent global emphasis on standards, accountability, and testing regimes. For those agitating in favor of reform for boys, standards and the tests that come with them are real, material concerns. That boys don't do as well on some tests has genuine effects, because, for students, such tests increasingly act as gatekeepers for higher education and, ultimately, middle-class lifestyles; for schools poor performance on tests can mean loss of money, students and staff.

Most of the concern about boys' achievement has regarded literacy. Again, this comes not from just Australia but from many nations. The Organization for Economic Cooperation and Development, a group representing 30 industrialized countries, has found that in every member nation fourth grade girls have a statistically significant advantage over boys in tested literacy.[3] For 15-year-olds, the gap stands even wider in all member countries. While there is a temptation to urge a biological or physiological interpretation of these cross-cultural results—that, since it happens pervasively, boys are "naturally" less suited to literacy because of their brains, testosterone, or whatever—one cannot overlook globalized masculinities that have in many ways standardized the socialization of males around the developed world.[4] In other words, it could be that boys in those countries are socialized in ways antithetical to tested literacy.

Headlines from around the world have raised alarms over a kind of *general* problem for schoolboys. Many suggest that boys are not suited to schools' "environments," with some arguing that schools are female-oriented and boys' proclivities and abilities—even their brain structure—are devalued or overlooked.[5] Explanations abound, particularly regarding boys' achievement in English, including lack of emotional skills and vocabulary, prudish teachers who don't appreciate boys' tastes for violence and scatological humor, social pressure not to do well in a "girly" or "gay" subject, and a rigid, narrow curriculum that privileges school literacies over out-of-school literacies.[6] This firestorm of disadvantages worldwide, some say, has led to more and more boys leaving school early and gaining fewer higher education degrees than girls.

Even with the heated attention on boys' academic fates, somewhat less attention has focused on the second major area: boys' social, personal, and economic conditions. This is partly due to an almost myopic focus on academics as conservative tendencies push "back-to-basics" approaches that reduce the social influence and mission of schools. Nevertheless, reform efforts have not squelched the public's worry over the social conditions and outcomes of schooling. Why else would conservative political movements focus so much attention on schools? Also, practitioners must confront many social problems to successfully educate students amidst remarkably withered social services in many countries.

Violence, for example, has been of great concern regarding school-age boys, with the decidedly male character of school violence being well established.[7] In nearly every human society, boys and men are disproportionately the perpetrators and the victims of violent crime. Most educators similarly admit that boys dominate in disciplinary referrals, suspensions, and expulsions. These facts may tend to hide other important dynamics relevant to gendered violence and misbehavior, though. Race, for example, has also become a central dynamic in the analysis of masculinity and school perceptions of who is or isn't a danger or nuisance.[8] Such perceptions often trap students of color into cycles of disciplinary trouble that are hard to escape, even in all-black, all-male schools.[9]

Bullying, a related topic, has also received much attention worldwide. School shootings, especially in the United States, have propelled this issue because bullying has taken much blame for these explosive events.[10] Vast practitioner effort, often implicitly focused on boys, has sought ways to decrease bullying.[11] Bullying, for many, remains the center of gravity in a constellation of social ills affecting boys, from homophobia to depression and suicide.

Many other social ills are attributed more to boys, as well. Drug and alcohol abuse, along with drunk and otherwise dangerous driving, has sparked concern. General risk-taking behavior has similarly been targeted, as are greater rates of male suicide, depression, and other mental health concerns.[12]

Economic worries, too, have followed boys. In many countries, of course, men hold most of the high-profile jobs in finance, technology, industry, and government, and men continue to earn, on average, more than women for the same work. Nevertheless, unemployment often affects men more, men's relative wages are decreasing (see chapter 6), and many "first world" economies are seeing marked declines in traditional sectors of male employment, such as manufacturing. Also a major issue, the lower percentage of males gaining university degrees tends to invoke economic fears since high-paying jobs increasingly require such degrees.

Suggested "fixes" for boys' problems, both academic and socioeconomic, have been nearly as plentiful as the problems themselves. Many fall under a "tips for teachers" rubric.[13] These present too simple fixes for highly complex problems, such as improving boys' reading ability by avoiding fluorescent lights.[14] Others, typically with more nuance, involve therapeutic interventions, sometimes, for example, promising help for boys in communicating or in overcoming behavioral disorders, criminality, or drug problems.[15] Still other solutions, some progressive, have been suggested that represent difficult, sometimes structural, alterations.

Perhaps the most frequently suggested "fix" is the worldwide push to increase the number of male teachers. From Iceland to Australia to the United States, many have called for increasing the number of male teachers to ameliorate boys' problems.[16] Largely, these calls provide little or no empirical evidence of the male teachers' abilities to be role models for boys or to teach boys better than female teachers. Still, the calls persist. While frequently rather unabashed attacks on female teachers and single mothers, calls for male teachers also often overlook dangers presented by any simple body count of males, for male teachers don't always provide positive environments.[17] In addition, many schemes designed to attract male teachers are woefully inadequate, and they often overlook the vastly complicated reasons males avoid, or leave, teaching.[18]

Interest in research on boys' issues has not involved concerns of only the public, social services, and practitioners. Rather, interest has also come from academic research. This has included ethnographies of boys' subcultures, typologies of masculinities, and explorations of diverse and marginal masculinities.[19] Others have sought to connect racial and social class dynamics with masculinity.[20] Still others have sought to establish the role of masculinity in schools using lenses of linguistics, history, and critical literacy, among others.[21]

Feminist and profeminist critiques of, and arguments against, the concerns about boys have also demanded much attention, and for good reason.[22] Many of these scholars have successfully countered the question "what about the boys?" with "which boys?" in an attempt to disaggregate the racial, religious, regional, linguistic, sexual, and class influence on "boys'" achievement and outcomes. They have effectively argued against viewing *all* boys as at risk when in fact only specific boys are.

These many issues, and their opposing views, have become standard to any debate over education, even among those who rarely pay attention to education. How has this happened? How has a particular, largely conservative set of issues become a common sense?[23] I argue that numerous dynamics have created conditions ripe for the growth of boys' education debates and, more important, boys' education *policies*.

Etiologies of the "Boy Turn"

Rather than coming "out of the blue," as some have previously asserted, numerous identifiable factors have contributed to the turn to boys.[24] First, and perhaps most influential, a group of highly accessible and well-publicized popular-rhetorical books on boys' education have captured wide attention. Books such as William Pollack's bestselling *Real Boys* and Christina Hoff Sommers' *The War Against Boys*, for example, have reached millions with warnings of increasing psychological and social harm being done to boys in modern society. Steve Biddulph, a bestselling Australian author, makes similar claims.[25] In general, bestselling books like these base their arguments on the existence of a "battle of the sexes" (note martial terms like "war" against boys), on biological determinism, and on a notion that boys have a "toxic," self-harming gender role. Each has gotten a greater audience with such claims than perhaps any theoretically oriented author writing from the feminist or profeminist perspective.[26] The popular-rhetorical books, indeed, have been heavily involved in the globalized flows of boys' education knowledge worldwide.

Media attention to boys' issues also work in this popular-rhetorical tradition. Several high-profile news events have contributed to this popular focus, commanding the media to ask serious questions about boys. In the United States alone, the "Spur Posse" incident of the mid-1990s in which a group of boys scored "points" for having sex with underage girls, the controversies surrounding the entrance of women to the Citadel military college and the Virginia Military Institute, and, again, the series of school shootings epitomized by the 1999 massacre at Columbine High School and the 2007 Virginia Tech University rampage have all placed boys, their socialization, and questions of power, privilege, and violence in the spotlight.[27] These events have aided the moral panic over boys, and interventions have grown out of targeting such high-profile events.

The second major etiology for the boy turn, somewhat ironically, has been feminist theory (and the men's movement's theorizing based on feminist theory). Work by feminists throughout the 1970s, 1980s, and early 1990s in establishing gender's effects on women's lives, particularly gender role theories, also opened the door for questioning the male role.[28] Those who examined the male role—whether educationalists, mythopoetic writers, social and religious movements such as the Promise Keepers, or even antifeminists[29]—identified vital social issues needing examination and intervention. These included the familial, social, economic, spiritual, and physical impacts on men's lives from labor, emotional disconnection, health concerns, divorce and custody disputes, body image, and violence. These concerns have led to similar theorizing about boys' lives.

Third, and another unintentional feminist contribution to the boy turn, the original formulation of gender equity indicators is now being used to argue for males' disadvantage. Kenway and Willis term these initial indicators a "strategic mistake" by feminists.[30] By basing equality on enrolment and test score gaps rather than, say, the economic and social outcomes of education, the groundwork was laid for boys' advocates to claim disadvantage at the first sign of access or test score advantages for girls. This presents feminist reforms with a significant challenge, for, to change equity indicators now, thus excluding boys, may seem to the public—rather than an evolving realization of the nuances of gender's effects—to be self-serving and cynical, or at least "out of touch" within an international context of high-stakes testing and accountability.

A fourth major prompt for the boy turn has been the rise of the New Right—the conservative restoration since the 1980s' ascendance of Thatcher in the United Kingdom and Reagan in the United States—and neoliberal reforms in education. While many point out the New Right's explicit aims of backlash against women, the *structure* of their educational reforms, particularly the interconnected processes of privatization and accountability, have accomplished more than their antifeminist rhetoric ever could.[31] This has been particularly true in England, where neoliberal reforms have led to an educational choice structure in which schools compete with one another for students. Administrators and teachers are forced to overvalue test performance lest they lose students and, consequently, their schools or their jobs. What results in such a milieu is a method of what Gillborn and Youdell call "educational triage," in which limited resources are funneled to those "on the bubble" of passing tests.[32] The gendered implication is that, since boys dominate the lower score ranks, funding goes disproportionately to them, so advances in equalizing the curriculum for girls, particularly in language arts, may be rolled back to better suit boys. In this way, New Right educational reforms create a "structural backlash" that challenges feminist victories without having to engage in explicit antifeminist rhetoric.[33]

One shouldn't overlook, however, *explicit* backlash politics, the fifth catalyst of the boy turn. This comes in constant claims that girls have made sufficient strides educationally and in many ways have surpassed boys. The tenor of such claims varies, from explicit, virulent attacks on feminism to more "reasoned" debate about modifying the "feminine" nature of schooling. Perhaps the best scholarly work on this comes from Australia, where Lingard and Douglas have examined the structural, policy, and media backlashes spurred by the "what about the boys?" debate and where Kenway and Willis explored the backlashes that occur in schools, among teachers and gender equity coordinators, and between students.[34] In general, backlashes feed on anxieties, threatened beliefs, and self-interest, so gender, in its high visibility as an identity politics, has been a major source of such feelings since the 1960s.

A major anxiety produced and fueled by gender politics, and the sixth cause of the boy turn, are changes in the economy and the workforce. As many have argued, the economies of "developed" nations have seen a "feminization" of the workforce; that is, industrialized economies are mainly growing in the service sector, in jobs traditionally held largely by women.[35] Additionally, the workplace cultures in "new capitalism" increasingly value "feminine" modes of interaction, such as working in teams rather than as atomistic competitors.[36] In general, males have been largely unequipped for these changes, and, as Arnot, David, and Weiner argue, schools have failed to help boys make the transition:

> Young men have been expected to adapt to an increasingly unstable set of circumstances in the work sphere, threatening the conventional basis both of masculinity and its associated ideal of the male as breadwinner. Such instability has been deepened, we suggest, not by the work of schools challenging and transforming masculinity, but rather by their failure to do so. While schools challenged girls to adapt to new circumstances, young men were not offered similar possibilities to adapt to social and economic change, even though the restructuring of the workplace and the family called for men with modern and more flexible approaches to their role in society. New sets of values, aspirations and skills were being asked of men as workers, husbands and fathers. The failure in the last two decades of government, society and schools to address the prevailing forms of, and ideas about, masculinity, particularly in relation to changing work identities and challenges to the patriarchal dominance of the male breadwinner, has had negative repercussions for boys.[37]

What the authors really refer to is the worldwide "crisis of masculinity." While uses vary, "crisis of masculinity" commonly refers to perceptions that men act in ways harmful to themselves or others due to cultural, economic, or political conditions that prevent them from fulfilling a (culturally specific) "traditional," masculine role. For example, Susan Faludi, in her recent book *Stiffed*, describes a crisis of masculinity in the United States in which "broken promises" of patriarchal dividends, secure futures, and civic roles have created a masculine culture of lashing out, resulting in a rise in both domestic and public violence. Each context may involve different configurations of this process, but crisis masculinity across the globe has fed on (mostly young) men who are excluded from local economies; faced with the prospects of "doing worse" than previous generations; perhaps denied "full, waged citizenship in the nation-state"; and deskilled and displaced by the feminization of post-Fordist labor.[38] Cultural and political conditions have also contributed, particularly incursions by the United States' and other countries' militaries and cultures through globalization and neo-imperialism.

The crisis of masculinity causes concern not only among boys themselves, but also among their parents, which is the seventh mainspring of the "boy turn." Indeed, much of the impetus behind reform for boys comes from parents who are, with good reason, concerned about their children's futures. Importantly, this concern (at least its public face) has come largely from middle-class white parents, who, one could argue, feel threatened by the loss of dominance for their sons.[39] It is important, however, not to lose sight of the elements of "good sense" behind some of these concerns.[40] To ignore the sensibility of some arguments for attending to boys risks "pushing" parents who have valid concerns for the quality, safety, and outcomes of their sons' (and daughters') educations toward rightist positions, for conservative groups are willing to take parents seriously on this matter.[41] We shouldn't construe, however, conservative groups' willingness to address boys' issue as a sign that they are somehow "tainted," or inherently a rightist position unavailable to progressive groups. Rather, progressive ends may come from working with boys, like diminishing violence or expanding boys' emotional and cultural repertoires. In fact, many progressive groups—particularly good examples come from African American communities—are already working with boys, seeking to mollify their disadvantages (see chapter 6).

The final spur to the boy turn is the "thrill of the new" for researchers. As Maguire and Ball note, "There are clearly 'fashions' in research as elsewhere; some projects are more 'sexy' than others."[42] While at first glance perhaps an example of the frivolous self-involvement of the "ivory tower," the allure of a "hot" field wide open for new and "sexy" research draws much attention. Lynn Yates describes this allure nicely:

> What we found when we were looking at and attempting to interpret the tapes from our first round of interviews was that the boys in our study seemed interesting and our findings there "unexpected," whereas...we could find little [new] to say about the girls....we became aware that much of the feminist literature on schools with which we were familiar...did treat girls in sensitive detail, while leaving boys as a more shadowy "other."[43]

This "shadowy other" has great power over the academic imagination. We shouldn't ignore, though, the political economy of such research decisions. For scholars desiring to establish themselves (myself included), finding a niche within a "new" topic or extending a hot debate is a powerful draw. Publisher demands for marketable products, in turn, support this impulse, a fact not lost on those who must "publish or perish." The fact that this circuit of production influences the knowledge that is ultimately attained, and therefore the issues that are promoted or targeted in schools, should be of grave concern to social justice movements in education, for, as I show throughout, this powerfully shapes or limits research, funding, and policymaking.[44]

The "Boy Turn" and Australia's Policy Ecology

In the Introduction, I discussed how policies spring from a context—not only a local context, but also an extended, globalized context. For boys' education, the international-level ecology has historically had little to do with formal, de jure policy. Australia changed all that. In fact, Australia's policy, *BGIR,* has already begun getting international attention, as I noted.[45] That does not mean, however, that boys' education discourse flows only *away* from the Australian policy ecology. Australia *takes* a great deal from international sources, too. This includes relying on cognate countries for much educational research, especially England and the United States, for Australia's relatively small population precludes a research infrastructure to rival those other countries. At a more informal level, though, large flows of media and popular-rhetorical works influence Australian debates too.[46]

Some instances of globalized flows of discourse intersecting with boys' education are easy to see. *BGIR,* for example, cites the Spalding Method (a U.S. program; 115) and Reading Recovery (created in New Zealand; 114, 121) as fixes for boys' literacy difficulties.[47] Other examples are subtler. For one, a principal cited frequently in *BGIR* told me that Pollack's *Real Boys* is his "bible" for boys' issues.[48] While the principal never explicitly mentions him in his testimony, Pollack's influence is clear in the principal's words quoted in *BGIR.*

Another subtle example comes from the most vocal and perhaps influential Committee member, Rod Sawford, the Committee's deputy chair and senior Labor member. In speaking about the failings of educational gender equity research and policy in Australia, he says,

> In fact, one of the criticisms...has been that you can always trace them back to Dr Carol Gilligan's book *The Different Voice* [sic]....That goes back to the 1980s and yet journalists chased Carol [Gilligan] right around

the world for the quantitative research that she said the book was based on. Of course there was none. What there was, which has now become public, did not suit her argument and so she did not use it.[49]

Sawford clearly understands global flows of discourse, for he (albeit incorrectly) cites a major work of U.S. scholarship that has indeed been influential: Gilligan's *In a Different Voice*. Sawford does not, however, speak accurately when he says that Australia's gender equity programs are based on this work. The *Gender Equity Framework* (*GEF*), Australia's national policy on gender in education, which Sawford criticizes elsewhere, never cites Gilligan, and no mention is made of her book's focus on moral development.[50] While Gilligan's findings may be foundational to theories of social construction like those in the *GEF*, Sawford clearly overstates its reach in actual, on-the-ground policy. More troubling, though, Sawford fails to cite the real source of his own methodological critique. This argument isn't, in fact, the work of "journalists" "chasing" Gilligan "around the world." Rather, it comes directly from Christina Hoff Sommer's antifeminist broadside, *The War Against Boys*.[51] Sommers alone tried calling Gilligan for the data nearly 20 years after the book's publication. That Sommer's book should make its way from the United States through global flows to a key politician in Australia seems both impressive and frightening. That he repeats it inaccurately as factual truth without citing the source—a single advocacy writer—is particularly disturbing. That Sawford represents the more *progressive* Labor party as its highest-ranking Committee member diminishes hope.

This is not the sum total of influence on Australia from the international context. Indeed, most of the internationalized etiologies I described earlier also operate in Australia. Its educational system similarly struggles with the devolution of manufacturing and social service infrastructures, economic inequalities between genders, and backlashes both explicit and structural.[52] The media, to a greater degree than many places, publicizes boys' education events, and research has been pervasive on the subject. Many, from politicians, such as former Australian Labor leader Mark Latham (discussed in the next chapter), to sociologists, believe that Australia exists in a state of "masculinity crisis."[53] It should be no surprise that Australia demonstrates the same tendencies as demonstrated by much of the international community, for Australia is part of that community. That the international ecology has turned to boys' issues makes the Australian push to frame national policy seem like a logical extension.

Conclusion

My exploration of the international inflections of the Australian policy environment isn't finished. I return to the international contours in the final chapter, where I explore the dynamics of the U.S. context that developed differently from the coherent, concerted way boys' education policy evolved in Australia. I also suggest implications for countries other than Australia at the end of each of the Australia-focused chapters. From the present chapter, though, it should be clear that the Australian boys' education ecology exists interdependently with a larger international milieu. Australia's boy debates have become loud and forceful, focused on the academic and the social, and driven by catalysts ranging from the economy to publishing. These catalysts push Australia in specific directions, but even these forces are not totalizing. Rather, they push up against sociocultural specificities, histories, and government structures and economies that bend and shape them in important ways and, ultimately, ensure that other contexts won't see exactly the same results. These cultural and social, historical, governmental, and economic specificities of Australia are the subject of the next chapter.

Chapter Two
Masculinity "Down Under": The Roots of Boys' Education Policy in Australia

Russel Ward's classic study of the roots of "the" Australian character, *The Australian Legend*, describes the mythical character of Australian men:

> According to the myth the typical Australian is a practical man [sic], rough and ready in his manners and quick to decry any appearance of affectation in others. He is a great improviser, ever willing to "have a go" at anything, but willing too to be content with a task done in a way that is "near enough." Though capable of great exertion in an emergency, he normally feels no impulse to work hard without good cause. He swears hard and consistently, gambles heavily and often, and drinks deeply on occasion. Though he is "the world's best confidence man," he is usually taciturn rather than talkative, one who endures stoically rather than one who acts busily. He is a "hard case," sceptical about the value of religion and of intellectual and cultural pursuits generally. He believes that Jack is not only as good as his master but, at least in principle, probably a good deal better, and so he is a great "knocker" of eminent people unless, as in the case of his sporting heroes, they are distinguished by physical prowess.... Yet he is very hospitable and, above all, will stick to his mates through thick and thin, even if he thinks they may be in the wrong.... He tends to be a rolling stone, highly suspect if he should chance to gather much moss. (1–2)

Ward ascribes the genesis of this character to bush culture and the shearers and range riders of the outback. This rough, manly culture made its way back to the cities dotting the coast—where the vast majority of Australians continue to live—via bush balladeers, the poets of the outback. While perhaps a stereotype or mythical idealization, this perception shapes Australian culture, even for many Australians themselves.

The perception travels the world, too: many Americans think of khaki-clad crocodile hunters and men of the outback when they think of Australians.

In this chapter, which lays out key contextual features of Australia, its culture, history, and education system, I want to avoid essentializing and mythologizing Australia. Too much diversity and subtlety exists in Australia to make sweeping pronouncements and wide-ranging typologies. This is particularly salient in any discussion of masculinity, a topic that has attracted scholars toward treating "hegemonic masculinity" as if it were a search for stereotypes in the media or for typologies of boys.[1] Still, masculinity usually operates *as if* ideals and typologies matter. The *materiality* of masculinity is hard to escape. Therefore, exploring dominant conceptions of masculinity within Australian culture—keeping in mind that place, participants, and history make such attempts partial and ephemeral—becomes a crucial task, particularly when policy—like *Boys: Getting It Right* (*BGIR*), I argue—gets written to shape or save masculinity.

This chapter explores more than Australian masculinity, though. To understand the genesis of *BGIR*, one must understand the history and context from which it springs. It didn't fall from the sky or appear under a rock. Certain individuals with particular responsibilities *created* it, struggling with specific constraints to serve identified ends that grow from unique histories including identifiable precursors to which it must respond. In this chapter and those that follow, I detail this context. Rather than being causative, I suggest that these factors and characteristics are contributory, placing limits of tradition, structure, and policy on what can happen now and into the future. Surely, any organization or nation can transcend the limitations of its history and culture, but it cannot start from nothing, tabula rasa (or *terra nullius!*).

Australian Characteristics Impacting *BGIR*

The Colonial Legacy of Morality

In the late 1770s, Americans started a war to seek their independence, and, succeeding, left England, their "mother country," with no place to put convicts. At around the same time, in 1770, Captain James Cook was "discovering" the great southern continent, later named Australia (from the Latin, *terra australis incognita*, meaning "hidden southern land"). Of course, Aboriginal peoples had already been on the continent around

60,000 years—or about 2,400 generations to European-Australians' current nine generations.[2] The Portuguese and Dutch had also seen the continent before the English, beginning in the 1600s, but it was the English who eventually—though not easily and never completely—"conquered" it. England began colonizing Australia as a prison colony in 1788.

Vestiges of being the British Empire's warehouse for criminals can be seen in Australia's culture and education today. Perhaps chief among these is education's constant moral savior role. According to Inglis, schooling's mission to reform "the rising generation" became official policy early, under William Bligh (of *Mutiny on the Bounty* infamy; his governorship in Australia ended the same mutinous way), who was charged by colonial authorities to "interfere on behalf of" the children of transported parents.[3] This role for education, as curative of moral and social ills, has continued throughout Australian history. Crotty, for example, details the many ways panics over boys led to massive changes to the educational programs in Australia around the turn of the twentieth century, particularly the advent of physical education in boys' schools, the rise of boys' adventure stories, and the growing role of boy rescue and youth movements.[4]

Indeed, this belief in schools as a means to improve morals still operates and shows up in *BGIR*. The policy grew from moral panic over boys' worsening academic and social indicators and backlash against what some perceive as feminism's attacks on traditional morality, a sense that boys will founder on the rocks of modern society and the new economy without urgent action. *BGIR*'s foreword makes this clear:

> As well as focussing on defined knowledge and skill objectives, [helping students achieve their potential] includes the development of attitudes and values which best equip them for life and for active and productive participation in society.
> Yet, the challenges of the classroom are becoming increasingly difficult. Social and economic change has impacted on societal expectations, student needs and attitudes, retention rates and educational policies and programs. Not all students or groups of students have fared equally well in our education system or with social change. (vii)

The slippage between morality, definitions of equality, and the purposes of schooling are complex here, but clearly concern for boys' moral states is central. Saving boys will save Australia.

Relationship to England

At this writing, Australia is still a British colony, with the true head of state being the Monarch of England. Of course, Australia's current relationship with the United States rivals its relationship with England, but its ties to England are in many ways deeper and more structural. While in practice Australia oversees its own affairs through its own Parliament, and while movements are afoot to sever ties to England (a republicanism measure nearly passed in 1999), this independence has mainly been illusory and partial. The best example remains the 1975 dissolution of the Whitlam Labor Government in Australia by Queen Elizabeth, a move that, while decried by many as interference in Australia's sovereignty, voters eventually ratified by giving the new Fraser Liberal Government an overwhelming majority. Of course, such control over (or interference in) Australian policy has been episodic and infrequent. Nevertheless, Australians have freely borrowed from Britain, imitated the British, and compared themselves to Britishers throughout their history. This has led to Australians creating a similar government (discussed later), following similar educational trends, and frequently using English research and political thought in the Australian context.[5] Despite this historical allegiance, Australia, because of its location, habitat, and economy, has rarely blindly, without alteration, copied England's policies, politics, or educational system. Anything borrowed had to be recontextualized (in Bernstein's term) for the new nation.[6]

These same tensions of recontextualization appear in the creation of *BGIR*. England's own debates about boys certainly encouraged demands that the topic be addressed in Australia. The amount of debate and publishing on the topic in England during the mid- to late 1990s, when Australia's discussions were also nascent, exceeded other countries' efforts. Take the words of Richard Fletcher, a government-celebrated expert on boys' issues (see chapters 3 and 4), during an Inquiry hearing:

> No, I do not honestly [know of other countries where gender differences are opposite Australia's], but I am not all that familiar with the international literature. Like you, *I have seen the literature mainly from the UK*. Recently we got some from the Netherlands and the US. (emphasis added)[7]

Even for others not explicitly admitting it, a major source of information to the Committee was England; there are 40 separate references to boys' education in the United Kingdom that appear in the *Hansards*.

The point isn't so much that Australia borrows from other countries. All countries borrow. Rather, Australia's borrowing most strongly comes from *particular* countries that Australia is, or wishes to be, like.

Size, Population, and Geography

Many aspects of Australia make it distinctive from other so-called Western, industrialized countries and thus require special consideration when developing (or imitating) policy. Chief among these is the continent's size, its relatively small population, and other factors of physical and political geography.

As figure 2.1 shows, Australia consists of six states (New South Wales, Queensland, South Australia, Tasmania, Victoria, and Western Australia) and two major territories (Northern Territory and Australian Capital Territory), with other minor territories consisting of islands off the coast. These political boundaries have been stable since shortly after federation in1901, and they are significant principally because Australia uses a federalist form of government (discussed below), and, in their separation of powers, states and territories rather than the Commonwealth have dominion over education.

With only six states, the provision of education remains heavily centralized, but still largely reliable. Though Commonwealth control over

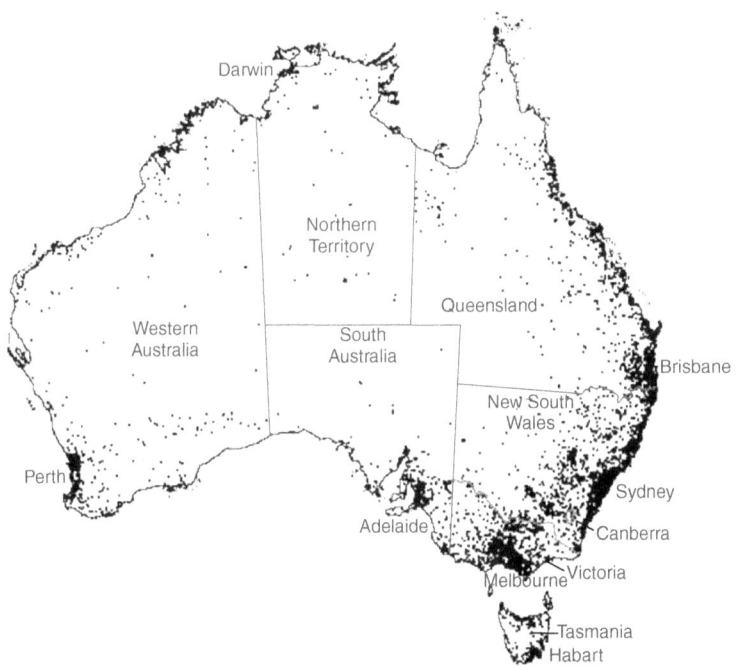

Figure 2.1 The Political Map of Australia, Including Locations of Capital Cities and the 2000 Population Distribution

Source: Adapted by permission from Australian Bureau of Statistics, "Population Distribution—2001 (Map)," http://www.abs.gov.au/Ausstats/abs@.nsf/Lookup/361F400BCE3AB8ACCA256CAE0053FA4 [accessed January 16, 2008]. ABS data used with permission from the Australian Bureau of Statistics (www.abs.gov.au).

education has grown dramatically since the 1970s, the states still dominate educational provision, and intergovernmental groups that negotiate joint concerns, especially the Ministerial Council on Employment, Education, Training and Youth Affairs (MCEETYA), also mediate the Commonwealth's influence.[8] This may be changing, though; see chapter 4.

Another factor becomes obvious when looking at Australia's political boundaries: each state has to cover a massive region with their educational services. Australia is about the size of the continental United States, but as of 2001 the country had fewer than 20 million inhabitants compared to the United States' 290 million. More than 85 percent of Australia's population, however, can be considered urban or suburban, and these urban/suburban centers take up only about 1 percent of the landmass, area that hugs the coastline (also evident in the population distribution, figure 2.1). This is mainly due to the lack of arable land in the interior of the continent, what some refer to—*sometimes* acerbically—as Australia's "dead heart." Nevertheless, there exists a small and widespread rural population, a situation that has produced numerous educational innovations, such as the School of the Air, one of the first live distance education systems worldwide.[9] It has also caused difficulties, like the implementation of mandatory bush teaching for new teacher licensees.

Concern for rural and remote communities is not completely absent from *BGIR*, with mention of particular problems of rural and remote boys being raised in submissions, testimony (discussed in no less than twelve hearings), and the report itself (e.g., 14, 143). Several initiatives for rural schools were also funded by the Commonwealth through the Boys' Education Lighthouse Schools program.[10]

Race, Fear, Immigration, Population

Throughout Australian history, the impulse to explore and colonize the vast interior of the continent has inflected racial dynamics in profound ways. The treatment of Aboriginal populations deteriorated progressively as the desire to move inland increased. More recent successes by Aboriginals and Torres Strait Islanders in winning land rights (for example, the Mabo court decision of 1992) have exacerbated the considerable racial tensions already existing, with particularly conservative segments of the Anglo-Australian population believing that the Indigenous population is getting an unfair handout and/or are being handed over white heritage areas. (The irony should be apparent). Regardless, much of the Aboriginal population continues to live in rural and remote communities, and the provision of education to them has proven to be even more challenging, thus compounding already low achievement in schooling for them.

The sparseness of the Australian population has also inflected relations with Asians and Asian-Australians historically. The so-called yellow peril grew from a fear that Australia, because of particularly low population density in the northern areas, was vulnerable to Asian marauders coming south. During World War II this impression was heightened by Japanese bombings of Darwin (Northern Territory) and several towns in Queensland.[11] Such fears still circulate, with debates over immigration often centering on the largely exaggerated threat posed by a "flood" of Southeast Asian and Indonesian refugees overwhelming the continent's scarce resources.

Australia's relatively small population also has implications for the progress of educational policy debates. It tends to be easier to get educational debates discussed nationally in Australia than in other countries. Put simply, capturing and holding the attention of 20 million Australians is easier than doing the same with the United States' 290 million. The smaller-sized media for this smaller population creates less of a din of information, too, and makes getting one's story heard somewhat easier. This has certainly helped the progress of boys' education issues.

Private, Religious, and Single-Sex Schools

Another critical context affecting *BGIR* is Australia's history of private, often religious, and sometimes single-sex schools. In the early years of the colony, schools were wholly religious institutions, mainly Anglican. As historian Alan Barcan shows, denominational fights and shifting political sentiments in England eventually led Australia, by 1895, to have the British Empire's first free, compulsory, and secular education system, and government funding for private schools was ended.[12] There remained a large private religious school sector in Australia after this time, until in 1964 Parliament reinstituted Commonwealth funding for private schools. The nongovernment system has expanded ever since. As of 1999, there were 9590 schools in Australia, 6970 public (72.7 percent) and 2620 nongovernment (27.3 percent). Nongovernment schools serve about 30 percent of Australian students, and most are now Catholic.[13]

Public funding for private schools continues to be contentious. The 2004 federal election saw the subject resurface, with Mark Latham, then Labor leader, proposing means testing for public funding, thus threatening to cut the funding for the wealthiest private schools. Such arguments recycle frequently because stark inequalities in school funding and facilities exist between state schools and private schools.[14] Indeed, in my own fieldwork sites (see especially chapter 5), inequalities among schools—especially between private and public, but even between some public schools—were glaring. One school had a new state-of-the-art performing arts complex while another school struggled to purchase paper for classroom instruction. Both schools, despite their relative desperation, received government funding.

While *BGIR* stays out of the debates over government funding, debates on boys' education are deeply inflected with concerns over private schooling. Private schooling has benefited from panics over problems in boys' education and has been seen as a fix for them. These fixes, indeed, are intertwined. Because single-sex schooling, particularly, has been highlighted among fixes for boys' educational woes, some private (even privileged) single-sex schools have used the issue to advertise themselves to concerned parents.[15] Consider the advertisement of one elite, private all-boys school, Ipswich Grammar (Queensland):

> Embracing modern research into raising sons and educating boys, the school has established a Centre for Specialisation in Boys' Education. Research findings into the ways that boys learn best are integrated into all aspects of the School. For example, the School recognizes the role of the Arts and Sport in boys' education, as a way to express themselves, build self esteem [sic] and learn important life skills. Visiting specialists speak with students, parents and the wider community about issues relating to the education and development of young men.[16]

While perfectly reasonable to advertise successes in boys' education, such schools don't advertise that the boys most truly having trouble in education are generally not those who attend their schools. They also fail to advertise that certain all-boys schools (not necessarily this one; I have never visited it) are highly masculinist and are damaging to many boys who don't, for example, play on certain sports teams or seem "appropriately" masculine or heterosexual.

Despite such problems, all-boys schools or single-sex classes within coeducational schools are often posited as key solutions for boys' schooling difficulties. Reasons presented to the Committee range from more willing participation by the boys to the school's ability to focus more on boys' specific needs.[17] Many also argued against single-sex education or warned of its dangers, but at least some on the Committee was persuaded of the benefits of single-sex grouping. Still, *BGIR* (perhaps wisely) avoided taking a stand on the question of single-sex schooling and classes.

Gender Relations

Gender relations in Australia largely mirror the major patterns of gender relations in all other "Western" nations. Saunders and Evans, following the work of Gerson and Peiss, characterize these relations as falling between the dual social tendencies of domination and negotiation.[18] Put simply, women are often dominated by men, but they resist and negotiate to win greater rights and rewards. While such a view is perhaps overly political and economistic in its interpretations—boiling down life and history to political control and capital, ironically neglecting feminist demands to also consider "the personal"—this nevertheless fittingly characterizes public sphere relations between men and women throughout the history of European Australia. For *domination*, women in Australia, just as in the rest of Western society, have been subordinated to men both economically and socially. Many Australian women face violence by men in domestic relationships.[19] Domestic labor largely gets divided by sex, generally with women continuing to take on the double shift of domestic work and paid work, though economic conditions are now changing this arrangement in complex ways.[20] In the public sphere, men dominate corporate governance and high-status professions and are more likely to be employed full time. Males are more likely to be elected politicians, though Australia has more women in Federal Parliament than many other countries, and women's issues have seen varying levels of interest.[21]

The second, obverse dimension of gender relations, *negotiation*, has been somewhat more successful in Australia than in other countries. Indeed, Australia ranks tenth out of 58 countries in the World Economic Forum's rankings of nations on gender gaps in economic participation, economic opportunity, political empowerment, educational attainment, and health and well-being; by comparison, the Scandinavian countries rank the best and the United States ranks 17th.[22] In opposing men's domination of women, Australian women's movements, throughout the twentieth century and continuing to date, have won major victories on women's issues ranging from universal, government-funded child care to government maternity payments, to government-funded contraception and reproductive health services.[23] Much of this success has been achieved because of the influx of feminist women in federal and state bureaucracies, a group given the Australianism, "femocrats."[24] Because the bureaucracy in Australia has the primary responsibility of implementing policy, the presence of femocrats has both encouraged the passage of feminist and pro-woman legislation, such as the *National Policy for the Education of Girls* and other gender and education policies (discussed below), and has been able to resist regressive backlash portions of boys' education movements.[25]

Naturally, this description tells only a part of the story of gender relations in Australia. Gender relations are complex, never monolithic or universal, and describing them in so short and conflict-oriented a way can gloss over positive relations and progressive potentials.[26] Still, the twin themes of domination and negotiation do indeed weave themselves through the history of gender relations in Australia, and these also affected the production of *BGIR*. Certain facets of the boys' inquiry process mirror these tensions. Domination is present, as seen from the calls to roll back certain parts of the educational gender equity reform that have gone "too far."[27] Negotiation is seen, too, as in the (pro)feminist scholars' and femocrats' resistance (see also chapter 3).[28]

"The" Australian Brand of Masculinity

Constructions of masculinity are obviously part of gender relations. Such constructions influence how a culture perceives itself and how others see it. For Australia, this might be especially true. Rough-and-tumble constructions of masculinity are a significant factor in how non-Australians view Australia, with a good example being the movie *Crocodile Dundee* as an introduction of Australian culture to many Americans in the 1980s (also subsequent iterations like "Crocodile Hunter" Steve Irwin on U.S. television).[29]

Beliefs about masculinity and manhood are also highly constitutive of the culture's *own* belief about its identity. As Epstein and Johnson theorize, all industrialized cultures create nationalized identities, particularly nationalized sexualities. Through televisual and print media alongside the workings of the state and institutions (notably schools), they argue, "nationality functions as a kind of meta-narrative of identity. It puts all the citizens in their places; it defines who belongs and who doesn't. It 'nationalizes'—names and rewards as national—some groups; excludes and punishes others as foreign or alien."[30] Masculinity—indeed, gender—certainly is such a meta-narrative, for it stands as a powerful sorting mechanism, defining who may access patriarchal privilege, not just separating women from men but also men from other men.[31]

Caution must be taken, however, in making an argument for a "nationalized" masculinity, as the title of this section cautions. A singular, national conception of manhood and the behavior of men does not exist *in practice*. While certain traits are valued within a particular context, not every male (or female) will aspire to them, and many will actively resist them. The traits of masculinity are in fact fluid, changing as the needs of hegemonic groups dictate, shifting as new groups gain hegemony over others.[32]

Even with the shifting sands of masculinity ideals, the ethos of Australian masculinity has taken on a permanence and prominence that perhaps other cultures cannot claim. Just what does this entail? In addition to the highly cross-cultural Western conception of the masculine in what Goffman calls the arrangement between the sexes—which stereotypically holds (heterosexual, white) men to be stronger, more rational, better suited to particular work in both public and private spheres, and responsible for protecting women, among other things[33]—Australia has other dimensions to its conception of masculinity that deserve note.

Larrikinism and Opposition to Authority
Depending on who you ask, the "larrikin," a male (almost exclusively) who defies convention and decorum and "sticks it" to the uptight and old, is either the embodiment of the Australian joy for life or the shameful product of too much drink and too little respect and breeding. Nevertheless, the larrikin represents what Australians and others think of Australian men. That males should embrace this spirit of defiance has no small implications for schools and policy. Much of the current desire for policy on boys' issues is a reaction against what has come to be known as a culture of "cool to be a fool" among (especially teenage) boys.[34]

Mateship
The concept of "mateship" is perhaps more important to the sense of Australian masculinity than any other facet. "Mateship" isn't a simple synonym for "friendship." As Martino and Pallotta-Chiarolli explain it,

> The term "mateship"...is used to define the camaraderie (including being anti-authoritarian, drinking, smoking, gambling and chasing women) and friendship between men, particularly in times of hardship such as during wars and pioneering in the harsh Australian outback.... It rests on shared understandings and discourses of loyalty, trust, and support, as well as on the premise that it is strictly a homosocial bonding between men which denies any sexual intimacy. (62)

"Friendship," they explain, typically implies a more intimate, personal relationship than does "mate." For many boys, such intimacy is threatening, and they choose to have mates, not friends.

Mateship powerfully shapes the arguments of *BGIR*. The policy's claims about boys being under-skilled in communication and interpersonal skills, for instance, implies—though they use "peer group" instead of "mate"— that the constraints of mateship are a drag on boys' school performance.[35] The Committee also argues, conversely, that "It is possible to employ peer influence to exert a positive influence over boys' behaviour and engagement with learning and to counter the anti-achievement attitude which affects some schools" (131). Whatever the influence, whether causing or helping to solve boys' perceived difficulties, the Committee clearly articulates a notion that mateship holds a crucial place in the construction of boys' masculinity.

Sport

Athletics, according to McKay and colleagues, "is a significant component of Australian society—as pastime, industry, locus of political power, source of media content, subject of everyday speech, and marker of ethnic, racial, sexual and national identity."[36] It is particularly key to the production of Australian masculinity, for it is "a major means through which ascendant forms of masculinity are asserted, promoted, tested and defended against 'rival' articulations of masculinity and femininity."[37] As Martino and Pallotta-Chiarolli point out, the connection of sport and masculinity has particular implications for boys in schools. Boys use intra- and extramural sport and physical education courses as "a site for policing, regulating and reinforcing certain versions of masculinity by peers and school structures."[38]

In the current boys' education debates, the centrality of sport for boys is arising again. The Oak Flats High School Cluster, a phase one Boys' Education Lighthouse School (see chapter 4), for example, developed a sports program "as a way of harvesting responsibility, teamwork and fair play."[39] Gardner and Little suggest rough and tumble, physical play with boys, which they assert helps boys open up emotionally in their Footy, Beer and Girls discussion group.[40] Whatever the ultimate effect of such programs, sport in the current debate provides testament to the enduring power of the athletics–masculinity nexus within the Australian imagination.

Militarism

Australia does not have a huge military or compulsory military service, yet the military is central in Australian identity and masculinity. The exploits of World War I "diggers" are a foundational myth, asserting to successive generations the vaunted values of what Macintyre calls "sacred themes":

> baptism under fire in the pursuit of an unattainable objective, sacrifice, death and redemption through the living legacy of a nation come of age. [The digger legend] told of courage and stoicism in the ultimate test of mateship and thus converted a military defeat into a moral victory.[41]

For many, Australia itself was born in the trenches, for there it proved, through the highest proportion of casualties of any country, that its boys were both fit for the manly duty of war and loyal until the bitter end to the fatherland.[42] There has certainly been resistance to such militarism, but this has been less formative to the Australian psyche.[43]

The military–masculinity linkage has reappeared in contemporary debates over boys' education. Though not a major feature, military training has again been suggested as a solution to this latest iteration of the "boy problem." Fletcher, for example, recounts a school principal instituting an army cadet program to attract and engage boys; mostly girls joined, though, a fact the principal found embarrassing, so he put the only two boys in front "so it wouldn't look so bad"![44] Thus, while not a central plank in pushes for boys' education, military education and the lingering belief in military masculinity as the nation's savior doubtless contributes.

Compulsory Heterosexuality
That hegemonic masculinity requires compulsory heterosexuality is neither new nor specific to Australia. In most cultures, as Connell explains,

> Hegemonic masculinity is emphatically heterosexual, homosexual masculinities are subordinated. This subordination not only involves the oppression of homosexual boys and men, sometimes by violence, it also involves the informal policing of heterosexual boys and men. Homophobia, in the sense of cultural abuse of homosexuality and the fear of being thought homosexual, is an important mechanism of hegemony in gender relations.[45]

Australia proves no exception to this, and it holds no more or no less tolerant views of homosexual masculinities than the United States, England, or any other "Western" country. Yet, I include compulsory heterosexuality here because it is powerfully entwined with the other aspects of masculinity discussed thus far. Many of these aspects are premised on either legislated (in the case of the military) or informal (in all cases) exclusion of homosexuals, a fact demonstrated by the reaction when homosexuality gets explicitly confronted.

Australian homosexual men have been visible in important ways in recent years. Much of the climate shift has come from greater acceptance of the Sydney Gay and Lesbian Mardi Gras each February, one of the largest such festivals in the world. Also, a popular home renovation "reality" show, *The Block*, in its first season, 2003, featured four couples competing to renovate apartments at Bondi Beach.[46] One couple comprised openly gay men, Gavin and Warren ("Gav" and "Wazzo"). Though their participation sparked controversy and calls for cancellation, the show's popularity and the couple's many guest appearances on other shows are evidence for the acceptance they won. Though some of the appeal was due to Wazzo's honking laugh and the couple's predilection for renovating in their underwear, most commentary focused on the "normality" of the couple's relationship and interaction. They were a loving couple with a relationship resembling almost any other, including those between the other renovators they competed against: they were tender at times, on each other's nerves at other times. For many, this "ordinariness" was a revelation, a stereotype-shattering experience.

Sport, too, has been the site of a major disruption to sexuality norms, especially the "coming out as gay" of Ian Roberts, a championship rugby player renowned for his toughness.[47] That such a notion should provoke outrage proves the indelible connection between sexuality and the core tenets of masculinity.

For schools, the link between masculinity and sexuality is central to the experiences of boys and men.[48] During the Inquiry, the Committee heard much from (pro)feminist scholars, education unions, and state education departments about the destructive consequences of homophobia and heterosexism. Indeed, while many scholars have been skeptical of the *academic* panics over boys, issues of sexuality and homophobia—which fit the "which boys?" imperative outlined in chapter 1—have been convincing enough to provoke diverse advocacy in favor of programs for boys.

Despite voluminous testimony about the ill effects of homophobia, the Committee ignored sexuality in *BGIR*, using the word "homophobia" (60) only once, and then too only as part of a larger quotation from a submission—and not even as the subject of the quotation. Though some members agreed that homophobia creates some of the problems noted by the advocates of "what about the boys?" some members of the Committee were hostile to the notion. Mr. Sawford, for example, was argumentative with those who discussed homophobia; he discounted its importance by calling it a "peripheral" issue and even congratulated those who eschewed or ignored it.[49] His influence on the Committee makes it no surprise that homophobia was ignored in *BGIR*.

Conclusion

I reiterate here a caution about the uses and interpretations of this section. Many other Australian masculinities exist in opposition to those discussed above, many of which are also highly successful. I also don't impugn all the above traits of masculinity; indeed, much can be admired in some of them, like loyalty and healthiness. If such traits were directed in responsible and caring ways, boys and the rest of society would be well served. Taken to extremes, though, such traits can lead to violence, harmful competitiveness, criminality, and social ills. I am also not implying the simplistic notion that *BGIR* only aims to reinscribe the existing or idealized structure of masculinity. While in some ways it is guilty of this, the situation is more complicated. These maxims of masculinity are in the background of many of the ideologies asserted, but simultaneously much of *BGIR* asserts a need for boys to adapt to changed circumstances. Still, it is productive to view the cultural impulse that led to the Inquiry—a moral panic over "the boy problem"—as based in anxieties over boys not living up to the masculine mystique outlined above, or at least anxieties that the mystique was no longer paying out the "patriarchal dividend."[50]

Government and Policymaking Structure

To understand any policy, one must also understand the governmental structure that underpins and limits it: a government's normalcies, systems, and traditions. This does not mean that governments don't sometimes make up the rules as they go—they do—but even making it up has to be done in the context of, or in opposition to, the existing government regularities.

Federalism

The basic structure of governance in Australia has, since federation and partial independence from England in 1901, been multilayered, with powers distributed to the local, state, and Commonwealth levels. Federalism implies the specific distribution of these powers. *The Australian Constitution* delineates the powers reserved for the Commonwealth government, called exclusive powers, while the other, residual powers are left to the states and territories.

Imminently familiar to U.S. readers are the tensions that this separation of powers causes in the political sphere, what in the United States gets called the "states' rights" debate. Where as in the United States, federalism and states' rights get used ideologically for everything from repelling a nationalized curriculum to instituting abortion restrictions and justifying the U.S. Civil War, federalism in Australia is primarily used for blame passing or scapegoating. If something has gone wrong, state politicians complain that the Commonwealth hasn't given enough money; the Commonwealth blames the states for not wisely spending the money they did get.

Still, in many ways, federalism is a valuable and effective system. It grew out of federation-era concerns that a single continent-wide government could wipe away the distinctiveness of individual states or overrun the rights of smaller and less populated states. Largely this has been successful, but there are many other benefits too, including enabling citizens to go "forum shopping" in case one level of government refuses their complaint, ensuring a state level that is more responsive to unique local needs and demanding closer cooperation between state and Commonwealth governments.[51]

The positive and negative aspects of federalism have fingerprints throughout *BGIR* and in the Government response. In the latter, Minister Nelson and the Department of Education, Science and Training (DEST) are careful to point out the recommendations that can be taken up by them and the ones that must be decided by the states, for education is a responsibility of state governments. Minister Nelson and DEST, though, also point out that they will use pressure on MCEETYA, the body of state and Commonwealth education representatives that cooperate on national education issues, to try to push the recommendations with which the Government agrees. In *BGIR*, too, the ingrained suspicions of the Commonwealth regarding the states not spending money as the Commonwealth wishes encourages them to recommend that states be subject to federal oversight (recommendation 24) and that the states be more consistent and comparable among one another (recommendation 23).

The Parliamentary System

As I pointed out earlier, Australia's government was largely inherited from the British. It is a Westminster parliamentary system with a lower and upper house. Yet, because small states were worried they would be dominated by the larger states, the framers of the *Australian Constitution* looked to the government structure of the United States. The lower house became the House of Representatives, which bases its proportions on population, and the upper house became the Senate, which bases its proportion on equal representation for every state.

As in the Westminster system, the House of Representatives determines who will be in the executive branch, for the House's majority party (the capital of Government) elects the prime minister and he (or she, though this has yet to happen) chooses ministers to head the various bureaucracies in specific policy areas (called "portfolios"), such as treasury, defense, and education. The party in the minority, called the "Opposition," chooses a leader of the Opposition and he (or someday she) chooses the shadow ministers, those who advocate the Opposition's counter-policies and criticize the Government ministers holding corresponding portfolios.

A ministers' authority to mold policy within their portfolio has a system of checks and balances to protect the governed. Foremost, and crucial to my analysis, the committee system provides a major means for information gathering and policy recommendation. Committees of various sorts conduct inquiries that gather submissions, hold public hearings, and produce recommendations that ultimately must be approved or rejected by the ministry. The ministers or their ministries may reject the recommendations outright, but doing so is rare because committees are largely seen as objective since they have both majority and opposition members.

With the ministerial system comes a large bureaucracy with responsibility for implementing policy and performing checks and balances over ministerial authority.[52] The bureaucracy, as mentioned earlier, has been a key site of struggle for femocrats, especially since women have only rarely been amongst the politicians. This has, for many years, been a solid strategy because the bureaucracy is made mostly of permanent, nonpolitical employees. Still, because the majority party controls the House, dominates committees, and leads the bureaucracy, the ultimate policy direction is theirs.

Another important check on politicians' unfettered power comes from Australia's system of compulsory voting. Since all adults must vote or potentially face a fine, politicians must play more to centrist politics. This has tendencies to assert more conservative, status quo positions and to weaken progressive or redistributive movements, but it also puts social issues such as women's concerns and labor rights on the agenda because oppressed populations are not discouraged from voting.

From this system of government has sprung several highly progressive policies on gender issues, particularly girls' education. These policies have responded to the cultural milieu outlined above and have prospered both because of and despite governmental structures that allow extensive policy participation by multiple layers of government, for the states, opposition parties (through committees), and bureaucracies where women could advance have all had roles to play, not as outsiders only, but as "inside agitators."[53] I turn to the fruit of this extensive participation by women next.

Australian Gender and Education Policy History

The genesis of girls' education policy began with the Whitlam Australian Labor Party (ALP) Government's creation of the Australian Schools Commission, the first body to bridge Commonwealth interests in education and state control over it. With the growth of awareness of social disadvantage because of women's and Indigenous rights movements, Commonwealth definitions of need began to shift toward numerous social oppressions, seen especially in the pivotal 1973 report *Schools in Australia*, the so-called Karmel report.[54] The oppression of girls and women was perhaps the most successful in catalyzing policy attention. Figure 2.2 provides a timeline of the progress of gender equity policy since then.

Girls, School and Society was the first of the reports specific to girls' education in Australia.[55] This important first step toward policy resulted from concerted struggle for recognition within the state by feminist groups. Largely the report focuses on liberal feminist explanations for unequal realities and outcomes of schooling, like differential treatment of boys and girls, differential expectations by teachers, sex role socialization that limits girls' choices, lowered self-esteem, and gendered segregation between and within schools. The principles for action the report offered, then, follow a model that similarly promotes "the capacity of both girls and boys to make considered choices" (157) without limitations imposed on them solely because of their sex. Facets to intervene in,

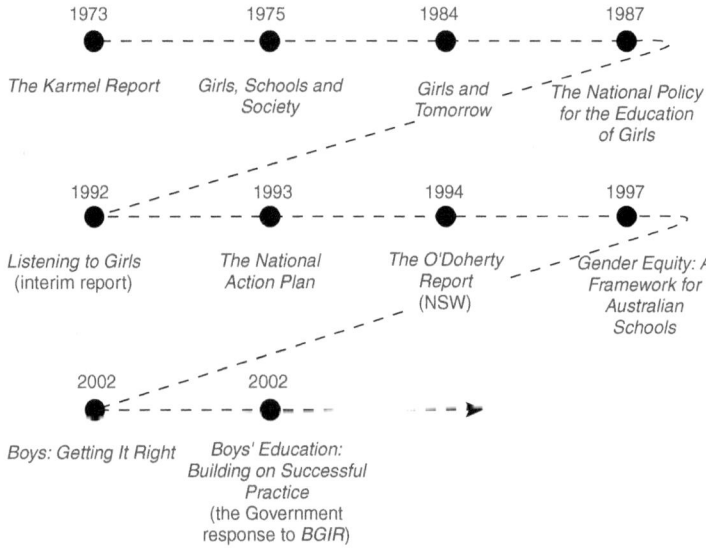

Figure 2.2 Timeline of Australian Gender Equity Policies, 1973–2002

thus, were the curriculum (both hidden and explicit), teacher training, pedagogy, promotion, vocational guidance, research, and opportunities to reenter the workforce after a period of absence. Rather than becoming official policy, though, the report only *acted* as a policy (much like *BGIR*) in the next nine years. As Daws pointed out, these years saw the states still taking primary responsibility for gender equity, creating policies of their own alongside Commonwealth funding for special projects and research.[56]

With the publication of *Girls and Tomorrow* in 1984, the practice of working solely at the state level began to lose credibility.[57] Based on a 1982 conference on girls' education, *Girls and Tomorrow* began a push in earnest toward national policy to replace the "limited and peripheral" (9) treatment girls' education concerns had received since *Girls, School and Society*. *Girls and Tomorrow*'s first recommendation was to create a national policy. In doing so, it also set out several recommendations that were more interventionist than its predecessors, recommending affirmative action to "feminise school hierarchies" (10) and the development of an action plan that would include a Commonwealth takeover of gender equity concerns.

With the ratification of *The National Policy for the Education of Girls in Australian Schools* in 1987, the first national policy on schooling was accomplished.[58] While this in many ways was an important development—as, for example, a case study in building organizational structures that can mediate the tensions of federalism within the education bureaucracy—its very existence highlights the tremendous successes of Australian femocrats. The policy was the result of struggles by key women, such as Senator Susan Ryan.[59] It demonstrates, most importantly, the key role of education and educationalists within the feminist project in Australia, and it demonstrates the visibility of gender within the cultural thought of Australians.

Notably, the policy spends much time outlining how a national policy might work within federalism. It stresses, indeed, that the framework relies on the belief that "meeting the educational rights and needs of girls is a responsibility of the nation as a whole" (10). Also, importantly, the *National Policy* seeks to establish girls' education reform within the realm of "professional responsibility" (28), perhaps to keep girls' education from being marginalized and to ensure participation from teacher preparation programs. In addition, the *National Policy* underscores the differences among girls, particularly indigeneity, class, language, disability, and geographic isolation. This recognition results from challenges from *within* feminism to the hegemony of white, middle-class, urban perspectives on gender equity. Another advance due *the National Policy* was the establishment of reporting and review procedures. From ratification on, the government was mandated to report annually during a five-year cycle, with a review and renewal process at the end the cycle.

The first of the subsequent five-year renewal cycles generated 1993's *The National Action Plan for the Education of Girls*.[60] This document addresses a perception that the *National Policy* was not practical in orientation, and that it provided too little guidance for schools to make changes. Thus, the writers of the *National Action Plan* (note the more practice-oriented title) describe their addition as "practical," "encourages direct action," "prepared for the use of people in the work places of girls' education," and a "document of hands-on assistance to schools, systems and their communities" (viii). Perhaps the most controversial and most illustrative of the theoretical changes taking place in gender studies and policy at that time was the first of their eight stated priorities: examining the construction of gender. A movement away from sex role theory was underway, and the take-up of social construction theories was perhaps best illustrated in the eventual widespread use of "gender construction" in the common parlance of Australian teachers. The term "gender" rather than "girls," however, has been interpreted by some as a moving away from the recognition of girls' particular sufferings based on gender, toward a watered down, even retrogressive, notion that boys suffer gender's downsides too.[61] Despite this newer wrinkle, most of the priorities outlined in the *National Action Plan* varied little from the initial calls for reform since *Girls, School and Society* nearly 20 years earlier.

What was indeed different in the context of the early 1990s, and perhaps partly responsible for the shift to using the term "gender" rather than "girls," was a growing movement to include boys' educational issues in examining gender. *Listening to Girls*, a 1992 interim report produced as part of reviewing the *National Policy* that eventually led to the *National Action Plan*, reported that "what about the boys?" was a frequently asked question during the Council's consultation process.[62] As Daws recounts, in the short term, calls for focus on boys were managed by including "gender" issues in the *National Action Plan* that might impact boys.[63] Eventually, though, boys' issues "swelled to fill" the temporary "vacuum" created by changes in the Commonwealth–state bodies in charge of gender equity policy—the disbanding of the National Advisory Committee on the Education of Girls and the creation of MCEETYA.[64] In turn, the terms of reference for the new Gender Equity Taskforce that MCEETYA created included explicit examination of boys' issues.

One result was the first-ever *state*-level parliamentary inquiry into the education of boys in New South Wales in 1994. The report that resulted was *Challenges and Opportunities: A Discussion Paper*, most often (and hereafter) called *the O'Doherty Report*.[65] Though sometimes depicted as such, the *O'Doherty Report* does not *itself* represent a backlash against feminism, though the report acknowledges that some of the impetus behind its inquiry came from conservative groups that believed focusing on girls had caused an "imbalance," to boys' detriment (3, 11). Not only did the Advisory Committee reject such backlash thinking, but also, as Lingard points out, it

> rejected a call for targeted boys' policy and a construction of boys as the new disadvantaged and argued instead for a gender equity policy to include both boys and girls and worked with a relational construction of masculinities and femininities.[66]

Indeed the *O'Doherty Report* called for "a major change in the way the educational community deals with gender equity issues" (3) by seeking

> a change in the attitudes of men towards women.... [I]t is implicit that men also need to reassess attitudes about their own role before society-wide change will take place. Many of the changes in attitude that will benefit women must be made by men.
>
> These changes will also benefit men. The material presented in this report clearly identifies that boys need help. The way forward is to release boys from the educational and social constraints of narrow gender stereotypes. (10)

The report also stresses that boys' education issues are not in competition with but instead are "interrelated" with girls' issues. That the report struck such a moderate-progressive tone was a victory for women's groups and femocrats who were increasingly being put on the defensive.

Whatever its progressive successes, little doubt remains that the *O'Doherty Report* was a catalyst for more boys' education reforms in the "what about the boys?" vein. While articulating a somewhat (pro)feminist perspective, it also included a list of disadvantages for boys, validated calls for more male teachers, and articulated a discourse of boys' disadvantages as "genuine" (1), "parallel to" but different from that of girls (1), and an "equal danger" for the remaining problems of girls if ignored (3); all of these were taken as legitimating the need for stronger, boy-focused reform. Indeed, precisely that kind of reform was achieved in *BGIR*, for many of the restraints called for in the *O'Doherty Report*—separate girls' and boys' coordinators in schools, a singular gender equity policy that incorporates the needs of both genders, integration of gender construction across the curriculum, and a focus on targeting unattainable masculinity as the root of both boys' and girls' problems—disappeared in *Boys: Getting It Right*.

An intervenient stage occurred in this retreat from (pro)feminist concerns, though. *Gender Equity: A Framework for Australian Schools* (also called the *Gender Equity Framework*; hereafter, *GEF*) was the second five-year renewal (in 1997) of the national gender equity policy, the prevailing national policy during the Inquiry into Boys' Education, and the one roundly criticized and targeted for "recasting" within *BGIR*.[67] The subject matter of the *GEF*, the indicators used in it, and its "strategic directions" didn't vary significantly from the *National Action Plan* it replaced. Still, much had changed to the displeasure of the (pro)feminists who had spent their careers constructing the previous policies. Lingard, for example, views the *GEF* as

> a real compromise document...being even more polysemous than most policy statements. The policy text is also not school- and teacher-friendly and unlike the *National Policy* (1987) has not been accompanied by an implementation strategy and accountability and reporting demands upon the states and ultimately the schools. It has basically been an endgame document, which held off the worst elements of recuperative masculinist politics, but which also placed boys very firmly on the gender equity agenda....Most importantly, it operated within a frame of presumptive equality [for boys and girls]..., thus serving to reconstitute the position of girls in the agenda.[68]

Daws further points to the impact of the change in language from "girls' education" to "gender equity," a drastic increase of which is seen in *GEF*. She argues that "the effect is to convey the impression that there is no systematic advantage or disadvantage to either group in any of these issues."[69]

Even with these considerable drawbacks, it is still important to view the *GEF* as an attempt to hold off recuperative masculinity politics and maintain an integrated approach to boys' and girls' education, one that views gendered problems relationally rather than as a result of feminist denial; of neoconservative longing for "pure," "boy-friendly" pedagogies that have been "forgotten"; of obfuscation and butt-covering by teachers unions; or of the feminizing influence of boys having mostly female teachers. *BGIR*, however, changed all that once and for all.

BGIR builds upon the long history of policy efforts described above, both girls' education and boys' education efforts. Indeed, the Committee itself outlines the history of previous gender and education policymaking.[70] While they conclude that previous policy has achieved "a great deal" for girls, they're less impressed with policy attention on boys. More specifically, they criticize the *GEF*, saying that it "does not separately research and identify boys' needs and it sets boys' needs solely in the context of what still needs to be achieved for girls" (64). Related to this, the Committee critiques "the negative approach for boys implied in most of the current policy material—for example, about boys *not* being violent, *not* monopolizing space and equipment and *not* harassing girls and other boys" (68; emphasis mine). Finally, the Committee believes that the indicators of disadvantage have been too narrow, focusing solely on educational and employment indicators. They conclude that these indicators don't fully account for gendered disadvantages and suggest that using a "range" of indicators—"such as rates of attempted and completed suicide and self-harm, drug and alcohol abuse, petty crime, violent crime, rates of imprisonment and homelessness" (67)—will demonstrate "distinct gender patterns." In other words, using such indicators will show policymakers that males have troubles too. These general dissatisfactions with previous policy translate directly to the first recommendation *BGIR* makes: the recasting of the *GEF*. The Committee suggests (a) creating "parallel" or "joint and distinctive" education strategies, one for boys and one for girls; (b) using "positive terms" for boys' educational needs; (c) allowing for school and community input; and (d) using a broader range of indicators of social disadvantage.[71]

The separation of strategies and policies by gender called for in *BGIR*—and materialized in the tender to recast the *gender equity framework* (see chapter 4)—is the continuation of a long process of shifting girls' policy toward boys' policy. Some have seen this as an "endgame" in girls' policy, a closing off of the political viability for girls needs, as seen in the shifts from "girls" to the generic "gender."[72] Whether the move to separate girls' and boys' strategies signals this or not remains to be seen (I don't personally believe this to be so; see my Coda), but it does indicate a growing acceptability of regarding boys as victims of gender and of schooling. It also opens the way, by giving latitude and oversight to local schools and parents, to using essentialist and pseudoscientific biological differences discourses in schools—what Epstein and colleagues refer to as the "pity the poor boys," "failing schools failing boys," and "boys will be boys" discourses, respectively.[73] That the Committee members positioned themselves against the tide of policy history signals perhaps a sea change in policy and an abandonment of the notion that boys' and girls' needs are interrelated and co-causative.

Despite such changes, *BGIR* continues important traditions from previous gender and education policy. Most clearly, the indicators of disadvantage have changed little from 1975's *Girls, Schools and Society*. A look at that report's subjects corresponds almost perfectly with the indicators used in *BGIR*: changing social roles, school participation and retention, post-school participation, curriculum, subject choices, teacher attitudes, teachers as role models, vocational training. Both focus on largely the same things, though some concerns have been elided over the years as conservative forces have made certain topics off-limits, such as sex education. Here again, the legitimacy of the *BGIR* report-policy is appealed for using traditional subject categories from other documents that are called "policy." That boys' education advocates can continue to use such indicators, though, shows the importance of both the original policy efforts—how certain indicators have become common sense about schooling—and the actual social changes that have occurred throughout most Western nations in the past few decades. It demonstrates, in other words, that *liberal feminist notions of schooling have been remarkably successful, changing taken for granted notions of what schooling ought to achieve if it is to be "equal."*

The Current Ecology of Boys' Education in Australia

As with biological ecologies, policy ecologies too shift and change in ways that privilege the survival, and sometimes flourishing, of certain groups and movements over others. In some cases sudden events can occur that dramatically shift circumstances—a volcano's impact, say, on nature or a terrorist attack's impact on policy. More often, though, incremental changes cause an ecosystem to become dramatically different—the gradual disappearance of predators in nature or growing conservatism in a policy context.[74] *BGIR* has been helped by a bit of both sudden and incremental changes.

First, and perhaps most important, Australia has followed the general trend toward political conservatism seen worldwide, what Apple calls "conservative modernization," though Australia's shift toward rightist positions has less to do with religion than do similar shifts in the United States.[75] Australia's conservative victories have been won on fiscal grounds (tax breaks and government payments) more than through "culture wars" over abortion, homosexuality, or religion (though this appears to be changing). One should be careful, though, not to discount cultural issues, for ideas about what counts as masculine and feminine are powerful, and attempts to redefine these have created easy targets on which to displace other frustrations. Feminisms, Aboriginal rights, gay rights, and other social movements get scapegoated for negative changes because they work in the foreground of the public stage, whereas much of the true decision making by those actually in power occurs "backstage."[76]

Concerns over boys are economic, too, the worry being that young men will lack the jobs necessary to fulfill their masculine roles. Crime, poverty, fatherlessness, incarceration, substance abuse, domestic abuse, and despair all follow. All the deconstructing of masculinity in the world does not make these material social impacts a figment of parents' imaginations, though admittedly many of those leading the charge (mostly white and middle-class) are less likely to experience such effects.

I should reiterate that boys' education initiatives, despite their conservative associations, are not *inherently* conservative.[77] Progressive educators can be and are concerned about boys' issues. Nevertheless, objectively and empirically speaking, *the conservative interpretation of the boy debates is dominant, and conservatism has driven the direction of policy*. The general shift to the right in Australian politics has brought the traditional political left, the Australian Labor Party (ALP), decidedly more to the center—and in some ways to the right—on boys' education issues. Indeed, growing conservatism nationally has meant that the ALP has had to co-opt issues, such as boys' education, that make it appear more centrist.

In fact, the ALP has sometimes actively championed boys' education to present itself as friendly to the "aspirational" middle class. Mark Latham, Labor's candidate for prime minister in 2004, in the speech launching his campaign, said,

> Those who have suffered most from the decline of social and personal relationships tend to be boys. Their school retention rates lag well behind girls. Their literacy levels are lower. And in disproportionate numbers, they are the victims of drug overdoses, road trauma and youth suicide. Our boys are suffering from a crisis of masculinity.[78]

Headlines in national papers the next day, front page above the fold, proclaimed, "Latham Targets the Boy Crisis."[79] This was the final signal that no politicians would stand in the way of an interpretation of gender equity as being now about boys. Though the ALP would still, for example, resist changes to the sex discrimination laws, the battle was clearly shifted away from asking *whether* we should worry about boys to a battle for nuanced positions over *how* to most quickly and efficiently help the poor boys in crisis.

The final macroscale contextual shift that encouraged national attention to boys' education was the increasing role of the Commonwealth government in educational policy at all levels.[80] As noted above, gender equity was a mainspring of initial federal interventions in education, and this and other social justice initiatives have benefited from Commonwealth support. This is true for boys' education too. But for a clear settlement favoring Commonwealth intervention in pan-Australian issues and but for the Commonwealth being the major source of funding for the states, boys' education movements would have little recourse to the Commonwealth and would have scant reason to do so. As it is, such movements have had opportunities to put boys' education on the national agenda, and they've made full use of it. In turn, the Commonwealth's increasing control over tied funding (giving money in return for specific initiatives), contributions to conferences, use of industrial relations reforms to control higher education curriculum and faculty, increasing calls for a national curriculum that is enforced "invisibly" through testing and comparability between states, and pushes for changes to sex discrimination legislation have given boys' education a power and mandate that could hardly have been imagined ten years earlier, even by the most optimistic of those agitating for boy-specific reforms. Again, however, this process of federal control has been a glacial shift, one that has been growing since at least the 1970s, and not one solely invented for boys' education.

In addition to longer-term macro-level contexts, several immediate contextual factors ensured that the report would not only become policy but would become a central policy. The biggest factor was the change of education ministers. David Kemp, the minister who called for the boys' inquiry—then chaired by Brendan Nelson—was, a year into the Inquiry, replaced as minister by Nelson. Thus, Nelson, an avowed advocate for boys' education reform, one who, as chair, had shouldered the task of defending the Inquiry in public, was given the task of overseeing the increasingly powerful Department of Education, Science and Training (DEST). Not only did Minister Nelson ensure that the boys' education issue that he worked so hard on didn't fade away unnoticed, but he also devoted significant resources and attention to the issue—both policy attention and media opportunity attention.

Soon after, Julia Gillard (ALP), the Committee member most opposed to the proposed boys' reforms and most aligned with the (pro)feminist resistance, left the Committee to become a shadow minister.[81] Thus, in one fell swoop, the ministerial post was given to a staunch, conservative supporter, and the Committee lost its most critical opponent of backlash politics. The remaining Labor members were either supporters of boy reforms or ineffectual in resisting the reforms' direction.

Finally, *BGIR* falls, intentionally or not, on the five-year renewal date for the national policy on gender equity that was established, as I mentioned above, with passage of the *National Policy for the Education of Girls*. With problems in the *GEF* as a dominant issue in *BGIR* and with the minister being a boys' education advocate, little could have stopped the momentum toward separate policies for boys and girls (discussed further in chapter 4). While this could be a positive for girls' education—ensuring attention for girls without dilution from a combined "gender" policy—the separation has thus far acted as a sanctioned way to focus solely on boys' policy, and little discussion of any girls' policy has been held.

Altogether, the educational policy ecology in Australia has definitely taken a "boy turn." Boys' education is currently *dominant* and *ascendant* in Australia, thanks to the confluence of conditions culturally, geographically, historically, educationally, economically, governmentally, and politically. What has resulted from this "perfect storm" of ecological conditions is a gendered, ideological, and political document that shows in its margins the indelible stains of the masculinity, the history, and the culture that produced it. In the next chapter, that document, *BGIR*, along with its creation, its textual reality, and its positives and negatives, becomes the focus of discussion.

Implications for the United States and Elsewhere

As promised, I want to step back periodically and talk about how my analysis relates to other contexts besides Australia. When looking at the various places where there have been panics about boys' education, clearly there are contextual differences that come to the fore, and an analysis like the one in this chapter is crucial to understanding the dynamics that create, maintain, and change boys' education. Moral panics over boys or over any other issue, of course, depend on the geography, history, governmental structures, other policies, beliefs about masculinity (and class and race), and other factors.

I detail many of the differing factors for the United States in chapter 6, but there are key points to be made here. First, many of the driving forces behind debates about boys are the same around the globe. This is both a function of globalization and of the common colonialist origins of many cultures and governments around the world. It is not an accident that the fiercest struggles over boys have happened in Anglophone contexts, not just because they share a language, but because they also share similar long-standing codes of masculinity that have circulated through colonial imposition or voluntary participation in the Empire's colonial projects elsewhere. This does not mean a one-to-one transfer of masculinities from England to its colonies, of course, but many underlying tenets are the same; the United States may have turned cricket into baseball and rugby into football, for example, but the focus on athletics as central to masculinity is the same. Thus, many of the principles I have outlined for Australia in this chapter fit easily for other countries, and I encourage readers to consider their context with each of the headings above.[82] Does geography, for instance, have anything to do with the spread of concern about boys' education? What governmental infrastructures support or resist the growth of debates on boys? Again, I answer such questions in relation to the United States later.

Second, analysis in this chapter underscores the necessity of monitoring situations at many levels as they change. Events small and large, as I detail in the section on Australia's current ecology above, can alter debates rapidly. The change of a government minister or pressure from new grassroots groups can change the course of events almost overnight. In some countries more than in others, monitoring can be difficult. In Canada, for instance, where the federal government takes very little responsibility for education, leaving it instead to the provinces and territories, following the movement in boys' education is a major undertaking. It requires thinking through the contextual factors noted above province by province. This can make research fragmentary, such that we might understand masculinist discourses in Quebec and educational gender equity policies in Nova Scotia, but a pan-Canadian context is difficult to gauge.[83]

The United States is similar in the need to look at state-by-state contexts. The U.S. state Maine, for example, began an inquiry into boys, but various factors led to a refocusing of the taskforce to study "gender" broadly—the linguistic shift was, in this case, a boon to girls' education.[84] Understanding Maine's context is required to understand this shift. All contexts, though, still exist within larger contextual patterns, such as those identified in chapter 1, and Maine's story is no exception. Readers are encouraged to consider such factors—from the local to the international—in their own backyards. Have local politics created room for boys' education debates? Are resistance groups holding such debates back?

Conclusion

Thus far I have given the background to the international and Australian scene in boys' education. This is a complicated ecology. To make more concrete how specifically such factors have shaped the actual product that is *BGIR*, the next chapter examines the production of the policy. There, and in examining policy levers in chapter 4, the deeply political and ideological nature of the boys' education policy process in Australia will become apparent.

Chapter Three
Boys: Getting It Right: Inventing Boys through Policy

> *Policy is, in a sense, an assertion of voice, or a cannibalization of multiple voices. Policy influence is a struggle to be heard in an arena where only certain voices have legitimacy at any point in time.*
>
> —Stephen J. Ball,
> "Researching Inside the State," 112

> *Education policies reflect the politics of the times and illustrate, at any particular time and place, which groups have more power to influence the state in its allocation of values.*
>
> —Mary L. Smith, *Political Spectacle and the Fate of American Schools*, 8

The previous chapter in part outlined the government system in which *Boys: Getting It Right* (*BGIR*) was produced: Australia's parliamentary federalism. It is a system not unlike that of the United States, discussed in chapter 6. This chapter focuses more specifically on the Committee itself, particularly the committee system that governs information collection and policymaking. From there, I deal specifically with the Inquiry into the Education of Boys, describing its missions, conduct, conductors, political particularities, the products it collected and produced, and those who influenced or resisted it. Rather than simply document these readily apparent features, though, I also dig beneath the surface, trying to reconstruct, through a kind of archaeological process, its deeply political and ideological meanings and implications. That is, I examine from where, as Ball says above, the House Committee on Education and Training "cannibalized" the multiple voices heard during the Inquiry. These are the voices that gain the legitimacy of which Ball speaks, the ones that finally are heard in the "arena" of boys' policy, and the ones with "more power" to affect the "allocation of values," as Smith points out. Legitimacy, indeed, forms a crucial component of the analysis. I also explore the policy as a political document, one that demonstrates the highly conservative context of the boys' education debate and the growing overall conservatism of Australian politics. I end, again, with some word about implications for other countries.

The Committee System and the Standing Committee on Education and Training

As in the United States and other countries, the Australian government relies on numerous committees for essential legislative and oversight tasks. These committees investigate problems, oversee ministries and ministers, or develop recommendations for policy action; the action itself is then taken, or not taken, by the executive branch. This was the obvious route for boys' education concerns. When boys' flagging literacy and atrocious behavior were thought to need addressing at the national level, these concerns were taken up by the Standing Committee on Education and Training (or EDT, as it called itself; hereafter, the Committee).

The Process

The Committee's process in conducting the Inquiry (see the timeline, Appendix) typifies the inquiry process in the House of Representatives. The Committee was given its terms of reference—or its guidelines for what to investigate—by the then minister for education, science, and training, David Kemp, on March 21, 2000. The Committee was to:

- inquire into and report on the social, cultural and educational factors affecting the education of boys in Australian schools, particularly in relation to their literacy needs and socialization skills in the early and middle years of schooling; and
- the strategies that schools have adopted to help address these factors, those strategies that have been successful, and scope for their broader implementation or increased effectiveness. (*BGIR,* xi).

Clearly these terms were broad, as they reference the schooling of half the population. Even so, the Committee went farther, even commenting in *BGIR* on intergovernmental functioning and professional groups. There is, nevertheless, a framing—or, limiting—of the debate through these terms of reference. The focus was on the *problems* of boys, for literacy and socialization are specifically picked out among all of the possibilities for indicators. If they had focused on, say, math and the economic rewards from schooling, the tenor of the Inquiry may have been quite different. The second term of reference also focuses attention on the school as a locus of creating and fixing problems. Rather than enquiring about, say, the government's role in creating militaristic masculinities or what could be done to ensure the media showcase progressive forms of manhood, the Inquiry sought only what schools could do. Much of how these terms of reference are framed derived from pressure groups that pushed Minister Kemp toward concern over boys.

The stated mission of the Inquiry shows some of the original intent of its founding, but some of that intent was known only behind the scenes. Julia Gillard, discussed in the previous chapter, when I asked if she had any attachment to gender issues or if boys' education had come up by chance, explained:

> Well I suppose the push for boys' education really came from the Minister. Look, there had obviously been publicity from time to time and discussion, you know, publicly and obviously in the education community about differential achievement of boys and girls, so I suppose that its topicality was what suggested it to the Minister, and the fact it hadn't been sort of forensically looked at before.

Mr. Sawford tells a similar story of how the subject of boys' education came about, citing "pressure" on the minister to pursue boys' issues.[1] A few years earlier, again, New South Wales' Parliament had released the the *O'Doherty Report,* the media and advocacy groups were loud voices on boys' education, and several conferences had been held around Australia (see also chapter 4).[2] Politically speaking, tackling boys' education spoke well to the base of the minister's electoral support, for conservative groups were strongly behind it.

After getting its terms of reference, each inquiry must be advertised to give the public opportunity to write submissions and to attend or give testimony at public hearings. The Committee also specifically invited, by letter, submissions and testimony from schools and state education authorities as well as relevant academics, service providers, and organizations. While sometimes public participation can simply be a political spectacle, it is clear from interviews with Committee members and the report itself that the views of the public were actually considered, even if with varying degrees of respect for those views.[3] Indeed, an analysis of *which* views were most respected forms the backbone of this chapter.

Massive amounts of information and arguments were produced and collected during the Inquiry, encompassing three major categories: hearings, submissions, and exhibits. (A fourth category of secondary, published materials will be discussed later.) The public hearings generated 1459 total pages of *Hansard* transcripts. The written submissions totaled 1899 pages (excluding those that were confidential). The exhibits, which included clippings, reports, and other miscellanea attached to submissions or provided during hearings, took up several file boxes in the Committee secretariat's office. Of those exhibits that I personally scanned, photographed, or downloaded, encompassing selections from 90 of the 178 exhibits, there were 1636 pages; but hundreds, if not thousands, more pages not available to me or not copied by me were available to the Committee.[4]

Certainly, given the volume of information collected, familiarity with the material varied among Committee members. While the chair and deputy chair may have had extensive familiarity, other members had limited knowledge. As the Committee chairperson, Kerry Bartlett, told me

> ...every public hearing we had there are fairly extensive briefing notes [summaries of submissions prepared by the Committee's staff, including possible questions members could ask] for each hearing and obviously I've got to be familiar with that material. And I suppose the reality is,...the supplementary members [meaning regular members, not those listed as "supplementary" in *BGIR*] perhaps don't as exhaustively study the material because the onus is on the chair to lead the questioning...

This process of briefing that he refers to results in a well-informed secretariat staff, chair, and deputy chair, but ordinary members may depend on that distributed knowledge rather than their own. The implications for their voting and participation cannot be adequately assessed, but it raises questions about the independence of the process from the ideological and political interests of a handful of people, some elected and some not. It also puts into question whether certain Committee members had the requisite "threshold knowledges"[5]—a knowledge base about curriculum, pedagogy, assessment, policy, and even gender construction—for making such large-scale policy decisions.

Still, when all was said and done in the hearings, submissions, and exhibits, the Committee had to produce a report, *BGIR*. The finished product can best be thought of as a polyvocal document, with traces of Committee members' voices as well as those of submitters and witnesses through quotations. Crucially, though, the latter voices are only of those whom the Committee *allows* to speak. Though the hearing and submission process was markedly democratic, the allocation of legitimacy isn't. Ultimate authority in the report lies with the Committee, though this, of course, does not ensure acceptance of their judgment or implementation of their recommendations.

The actual composition of the text, as frequently occurs in governmental reports, fell to a staffer in the Committee secretariat, James Rees.

Mr. Rees, naturally, was not at liberty to write just anything. Instead, the Committee met numerous times to craft the report it wanted. Mr. Bartlett explained the process:

> Then... the Committee secretary paid by the Parliament put together the draft report after consultation with myself and the other members of the Committee in terms of which direction we wanted to go. Lot of liaison, lot of discussions between myself and him [Rees] in terms of, you know, the structure of the report, what sort of emphasis we wanted and so on. There was then a *long* process of discussion within the Committee with the draft. So we'd take each chapter at a time, we met for couple hours at a time and just sort of arguing the finer points of it. "Was the emphasis correct? Do we like the terminology? Did it really accurately reflect our concerns about particular issues?"... The disagreements that occurred during that discussion process were not along [political] party lines. They were just sort of just different preferences in terms of terminology or emphasis.

Another Committee member, Sid Sidebottom (ALP), in an interview explained the tone of those meetings from his perspective:

> We also spent a good deal of time in the draft and editing stages of the report. There's a very very strong commitment to the *cause,* if you like, of boys in education—not to forget that we're interested in young people or people in education—but there was a very strong dedication to the cause inside the Committee. And that was reflected in *painstaking* reflection on just about every paragraph. *But* of all my involvements in committee work in the Parliament... I would have to say it was the most *committed* response from a group of individuals that I have seen.... We did *not* want to see another report stuck on another shelf gathering dust. So in a sense I think there was an almost unconscious effort to *provoke* a response. There are established views on education and gender equity, we believed, that existed—or established views, *institutionalized* views—and I think we wanted to provoke a response, and we were determined to get that, without being irresponsible.

This issue being a "cause" that the members were "committed" to, one for which they were enthusiastically trying to "provoke" an activist response, raises concerns about the degree to which the Committee was open to having their a priori beliefs challenged and how much they were also "committed" to protecting the interests of girls and women.

Sidebottom increases these concerns by describing the tenor and representativeness of participation in the Committee backroom. I quote at length because of the importance of this participation to the finished report and the ultimate implications for policy from the report's conclusions.

MWH: So I get the sense from you that all the Committee members were on the same page as far as their opinion on the boys' education issue....

SIDEBOTTOM: Yes. [Pause as if searching for words]. Yes. It almost makes it sound like it's circular, [*laughing*] right.... Instead of having the political linear, this was a circle. There was a *view* expressed, or a sentiment expressed, that the gender equity thing had gone too far. That it was generally succeeding for females—yet I'm not sure that the evidence suggested that—but in the process the boys had been left out.... Anyway, we were on the same page, but it was a fairly large page. [*Laughs*] And we did differ. You know there were some [members] more expert in an understanding of how we learn than others.

MWH: So would that be the major difference that would have cropped up between members?

SIDEBOTTOM: Yes, it was *interesting*, in the *main*, I think the debate in the report was dominated by males. It wasn't— The female participation in the report at our level was not as great or as enthusiastic or indeed as [*inaudible*] as was the male. Now that may be just coincidence. I don't know.... Let's put it this way, when we [*laughing*] did the [*inaudible*] and the drafting, there seemed to be a hell of a lot more males in the room than females. But it may have been a workload issue, I don't know.

I asked if he thought that the gendered participation might have shaped the report in a way that might not have happened had there been more female participation.

SIDEBOTTOM: Well, to be fair, one—No, there were two females that I recall—you know it's some time now [since the drafting]—I must say [those females] were fairly traditional in their view that they thought boys were getting a raw deal. So I suppose in that sense they [the women] weren't any different than the thrust of the actual report. But the actual substantive work on the report happened to be done by males.

Though Sidebottom dismisses that crafting the report "*happened* to be done by males," the implications are worrisome, for an apparently small group of males with an advocacy stance crafted a report that has created millions of dollars of programs, refocused much research and practitioner effort, and caused the redrafting of an entire nation's gender equity policy. If Sidebottom's characterization is true, the authority allocating the values and legitimacy within *BGIR* was largely male, white, middle to upper class, and committed in advance to the notion that boys were in need.

The Committee's Composition

Certainly, the Committee's *process* represents a determinative structure in the outcome of the Inquiry. Just as important, though, is the *composition* of the Committee. Who are the people on it? What are their backgrounds? What political and ideological affiliations—not to mention identity affiliations like race, class, and gender—do they represent? Understanding such questions can give insight into the results of the Committee's work.

Consider the composition of the Committee by party. The majority party must, by rule, have a majority on committees and select the chair. As the government of the time (both in the 39th and 40th Parliaments—the span of the Inquiry) was controlled by Prime Minister John Howard's Liberal-National Coalition (the conservative, right-wing party), by rule the Liberals outnumbered Labor members on the Committee (six to five, six to four, and seven to five, at various times). Given the chair's power to direct the committee's work and the simple majority needed to make final decisions, this structural advantage is determinative to the outcome. While every Committee member I talked to rejected any notion that the Inquiry was "party political"—that is, partisan—party politics was not unimportant, which I explain further in the next section.

Obvious from tracking the membership of the Committee over time, there was a greater carryover of Coalition members in addition to their structural majority. Five of the seven Coalition members on the final Committee were also members at the start. In contrast, only two of the five Labor members stayed on, and one of those was only a "supplementary member." The other was Labor's Rod Sawford, the most vocal critic of teachers unions, education department bureaucrats, and (pro)feminist academics—truly a conservative Labor member. What's more, one of the five final Labor members (Albanese) was only on the Committee for two months before publication of the report and, by admission, attended no hearings.[6] This lack of continuity might suggest that Labor had a lack of experience and knowledge of this issue—for how much can one learn about so complex an issue in two months?—that Labor ceded the issue and signalled so by not putting a dedicated force on it, and/or that internal politics within Labor caused a high turnover between Parliaments.

More proof of the questionable knowledge level of members comes through considering attendance at public hearings, major venues through which members could learn from experts and practitioners, particularly since many likely relied on briefing notes rather than reading written submissions themselves. Quite simply, numerous members missed a great number of hearings. The average Committee member attended only 51 percent of hearings held during her or his tenure. The best attendance was from Dr. Nelson, who attended all ten hearings held before he left the chairpersonship. Of the seven who were

Committee members for all 32 hearings, Sawford (ALP), the deputy chair throughout, had the best attendance rate at 84 percent. The final Committee, 12 members altogether, were the ones being counted when someone claimed that the Committee "unanimously" approved the report. Their average attendance was 47 percent. On average, then, each came to his or her decisions after having attended just under 15 of 32 public hearings.

The hearings' locations also seemingly played a part in attendance levels, and this demonstrates some politicking. Non-Canberra meetings were attended by an average of only 4.8 members while Canberra meetings averaged 6.9 members. Three members of the Committee attended no hearings outside of Canberra, two of whom (Johnson and Plibersek) were members of the final Committee. Apparently, though, the draw of a hearing in a member's home state was more powerful than most; members attended 66 percent of hearings in their home states during their membership.

Using averages, though, disguises the lack of requisite threshold knowledge of the issues for many of the final Committee's members. Leaving out Mr. Sawford, the other 4 final Labor members—Albanese, Plibersek, Sidebottom, and Wilkie—combined attended only 17 hearings, and, again, Mr. Albanese attended none. For the Liberals, 4 final members—Gambaro, Johnson, Pearce, and Cadman—combined for only 21 total appearances at hearings. This puts into question the knowledge base these 4 Labor and 4 Coalition members (a full two-thirds of the final Committee) brought to the decision-making process—a combined attendance of 32 percent—and perhaps helps explain the dominance of a few "committed" members Sidebottom describes above. When contextualized by the report's potential impact on millions of school children and educators—for the report advocates reforming teacher education, school structures, hiring, scholarships, discipline, curriculum, data collection, and even pedagogy—the light attendance by many members is startling. Just as "threshold knowledge" on gender and education issues is required of teachers, one would hope that those making policy on it would have certain threshold knowledge. The analysis of attendance at hearings, alongside a lack of females in much of the decision making, however, puts the level of threshold knowledge in question. It also raises questions about whether the report was "party political."

Partisanship in the Committee

Despite Sawford's assertions that boys' education shouldn't be a "political event," the entire issue is *inherently and unavoidably political*.[7] The issue involves the ascription of victimhood, the apportioning of research attention, the endowment of significant funding, and the construction of a discourse about what is "true" of boys' education. Politicians may get or lose votes based on it. Official knowledge is being developed.[8] Pedagogies and assessment are being suggested and validated. *How could it not be political?*

Nevertheless, each member I spoke to categorically rejected any "party political" nature to the Committee's work. They proudly proclaimed that the report was a "unanimous" one, free from the ideological bickering that characterizes many parliamentary debates. As Mr. Bartlett said,

> That process, it wasn't very party political. It was really quite—There was a high degree of agreement. The fact that it's a unanimous report indicates that. There wasn't much disagreement at all in those hearings. We were pretty much on the same wavelength.

Mr. Sawford, saying that it was not partisan, also said "And I think that's a very heartening sign, because I think if you put partisanship into education, we're all diminished by that." How can this civility exist, though? How, in a tense era for politics, where political compromise seems rarer than ever, could there be a committee that, taking one of the most bitterly fought issues in education globally, has little to argue about?

Though the Committee members view the lack of partisanship as positive, I want to trouble this notion, for the lack of such debate— perhaps ironically—indicates worrying signs for the development of educational policy. While one would certainly hope that the interested parties are taking an "objective" look at evidence and making decisions based on it—thus avoiding partisanship for partisanship's sake—it signals, instead, a retreat from feminist allegiance by Labor, or at least an evacuation of positions within the state that have been, until recently, used by femocrats to block harm to girls. Members' positions might not conform to traditional party platforms, true, but they are *at their cores* ideological and political. Consensus in this case does not signal lack of politics; *it signals that the Left has come to take up the Right's position.* I honestly believe that the Committee's purpose was an attempt to do good for Australian students, and they did weed out more radical factions of the backlash. However, a clear shift away from feminist policy initiatives has occurred, and the Committee's process and positions are both a symptom and a cause of that. Sometimes party politics are *needed* to resist the movements away from social justice. Dissenting opinion breeds caution and reflection into political processes that are otherwise easily blinded by the ideologies of those with political might.

Arguing the Politics of Boys' Education

To demonstrate that the Inquiry was indeed political, even without partisan squabbling, I need to reconstruct the sources of ideas in *BGIR*. Just what ideologies were presented to the Committee? What ideologies did the Committee bring with them? In this section, I explore the various arguments made about, for, and against boys' education concerns. These were the stuff of the eventual report-policy.

BGIR, according to my content analysis, makes 399 separate arguments relevant to boys' education. This represents a tremendous culling by the Committee, for, again, the sources of information available were prodigious. Arguments from the public hearings, for example, according to my analysis, separately numbered 2294. Add to this the massive stacks of pages in the exhibits and written submissions, and the Committee's task was no small one: to select from the many facts, opinions, philosophies, fears, lies, distortions, and omissions to forge a coherent argument of its own. That this selection process had to happen indicates in yet another way that this process was deeply political, as any report process is. *Who* and *what* was listened to, who and what was not, and why are key questions to answer in the political excavation of this process.

Obviously, tracing every argument in *BGIR* would require a volume three times this size just for the arguments alone. The Committee was presented with a tremendous breadth of topics, from alternative schools to teacher education to reading instruction, even to the ravages of color blindness.[9] So instead of an exhaustive analysis, I present a subset of arguments that the Committee asserted in *BGIR* and trace those to their sources, whether explicit or unstated.

First, though, two caveats are in order. For one, whether the Committee is agreeing with or taking influence from its informants or whether the informants simply agree with or even take influence from the Committee is often difficult to determine. In most cases the Committee *asked* for certain information, and, ultimately, terms of reference guided what the Committee sought. Are the witnesses' comments, then, what they truly thought, or were they prompted to make certain arguments? Despite this ambiguity, hereafter I tend, for simplicity's sake, toward describing the submitters and witnesses as "agreeing with" the Committee.

Second and crucially, the Committee didn't simply have to contend with the finite set of documents and testimony they took. They, like any group of learners, had to deal with, for example, their own existing knowledge and beliefs (about gender, about the purposes of education, etc.), media reports, things they learned from talking with others, their experiences as parents, and books they read. Thus, it is too simple to look at the submissions, testimony, and exhibits to gain a comprehensive understanding of their sources. Really, Committee members had the whole globalized policy ecology to accept, reject, or modify to fit with their final report.

Public Hearings

In the first step of analysis, I cataloged all of the arguments relevant to boys' education from the public hearing transcripts, both those made by witnesses and by Committee members themselves. The initial coding netted 2342 arguments. A second pass cut the list down to 2294 unique arguments.[10]

My sorting of arguments into categories suggests the general tenor of what information the Committee was given and what they sought. The "teaching strategies and educational programs" category contains the most arguments as well as the most oft-presented argument: pointing out a specific effective program for boys. This is no surprise, for one of the Inquiry's terms of reference was to determine "the strategies which schools have adopted" to address boys' education (*BGIR*, xi). The many school visits and testimony from practitioners also influence this result. Since the Committee's other term of reference was to "inquire into and report on the social, cultural and educational factors affecting the education of boys" (*BGIR*, xi), the finding that 30 percent of the arguments fall into categories I named "social, behavioral, and medical"; "cultural, social, and economic changes"; and "academic achievement and learning," combined, is no surprise either.

Perhaps more surprising, and not a term of reference, my "politics and political arguments" category was the second most frequent. This category, as the label suggests, encompasses broad issues of resource allocation, legitimacy, and ideology. It includes any argument that suggests what should be funded, what groups should receive time and attention, who controls what and ignores what, or what the law or policy of the land should be and to whose benefit. Of course, all arguments are political in a certain sense, but I reserved this category only for those arguments that were explicit in their political orientation. Still, the fact that this category was the runner-up for most arguments that were made illustrates clearly a central tenet that this book argues: *boys' education is an inherently political topic, and it is a main site of struggle over the distribution of resources, the validation and endorsement of deeply held beliefs, and the ability of identified groups of people to participate in society's institutions.*

Written Submissions

To fathom the full political depths of the Committee's knowledge production about boys' education, I reanalyzed the public hearing arguments by cross-referencing them against the Committee's other sources of information. My analysis of the written submissions proceeded in a different way from that of the hearings. Both to compensate for the density of argumentation in written prose and to counterbalance the methods, I coded *BGIR* for arguments first and then looked for particular arguments in the written submissions, recording who made them. This allowed me to trace the conclusions and assertions of *BGIR* back to their origins, just as the hearing analyses allowed the reverse.

The analysis netted 399 arguments in *BGIR*, obviously unwieldy for a full analysis of the submissions. Thus, from the initial list of arguments in *BGIR*, after numerous passes I chose a list of 18 that were particularly contentious (table 3.1). These, along with the 24 specific recommendations made in *BGIR* (see table I.1)—42 codes altogether—I searched for in the written submissions. Again, I was looking for support, not disagreement; I wanted to know which sources the Committee validated.

In tabulating the agreement between the submission writers and the Committee, I developed a measure of "agreement score" that took the number of arguments and recommendations the submitters and the Committee had in common and divided by 42, the total possible items of agreement (18 arguments and 24 recommendations). For the 202 submissions to the Inquiry, the average agreement score was 5.22 percent. In real terms, that means the average submitter made between two and three arguments that were also made in the report.

To provide a check against the bias possible in selecting myself the 18 arguments for which I coded—in other words, to ensure I didn't just select the most conservative writers—I then compared the total agreement score on all 42 arguments with the agreement score for the 24 recommendations only. Would the same people, in other words, who agreed with the Committee on the recommendations also be in agreement with the arguments I selected and thus mitigate any notion of coding bias? In fact, agreement with the arguments I selected and the Committee's recommendations were, in general, well balanced, for, of the 28 submitters

Table 3.1 Reduced List of Arguments from *BGIR* Used in Submissions Analysis

1. The *Gender Equity Framework* (*GEF*) is lacking (e.g., it does not address boys, uses wrong indicators, couched in negative terms for boys)
2. Higher school retention rates have been bad for boys
3. Social, economic, and policy changes have affected boys more (or disadvantaged boys)
4. Boys respond more negatively to bad teaching or curriculum than do girls
5. Boys and girls have different learning styles
6. Assessment methods disadvantage boys
7. Phonics help boys; whole language hurts boys
8. Small class sizes are good for boys
9. Relationships between boys and their parents and teachers are key to their development and success
10. The gap between life inside school and outside school is harder for boys
11. Male teachers, role models, and fathers are needed and are good for boys; positive representations of boys in the media and curriculum are good
12. Education departments, teachers unions, and some academics understate or ignore boys' needs [chapter 1]
13. Disadvantages of one group shouldn't be used to downplay the needs of another group [chapter 2]
14. Boys need to broaden their labor market awareness and skills
15. Educators have "forgotten" basic pedagogy that is effective for boys [chapter 4]
16. Single-sex classes and schools are not the answer in themselves; it is what you do in those classes and schools that counts [chapter 4]
17. Computers can motivate boys or help them overcome disadvantages [chapter 5]
18. Educators need to build an environment that is "affirming" for boys (promotion of a boy-friendly school culture) [chapter 6]

Note: Arguments not found in the executive summary are noted above with their location in *BGIR* in brackets.

that had a total agreement score of one or more standard deviations above the mean, only four (14 percent) didn't also have one or more standard deviations above the mean agreement with the recommendations only. Thus, I assert that the arguments I chose were indeed representative of *BGIR*'s political tenor.

After collating all of the written submissions that agreed with the arguments and recommendations in *BGIR*, I did the same with the hearings (discussed above). The resulting cross-referenced roster of informants (shown in table 3.2), then, are those who have a high level of agreement with the Committee's positions.

Explicit Citations

To understand whom the Committee listened to also requires, obviously, examining the sources *explicitly* cited in the report's text. Overall, 565 citations were made in *BGIR*. Of note, the testimony received in hearings was most utilized, accounting for 35.4 percent of all citations made, with 47 percent of all witnesses being cited in *BGIR*. Clearly the members were committed to using information gained from this most democratic part of the process.

The major question is who was listened to *most* based on whom the Committee referenced in the text. To determine this, the means and standard deviations for each information source were determined, and lists of those witnesses, references, submitters, and exhibit providers who were at the mean or above were produced (included in table 3.2).

The explicit citation counts also help to underscore the value of the "agreement score," described above. The higher the agreement score—again, the more the submitter's arguments coincided with the conclusions of *BGIR*—the more likely the submitter was to be explicitly cited in the report. Only 20.7 percent of those below the mean in agreement score were explicitly cited in the report, while 53.6 percent of those with a standard deviation or more above the mean in agreement score were cited explicitly. This makes perfect sense, for in composing its arguments the Committee would surely rely more on sources they strongly agreed with and that agreed with them.

Witness Time Allotment

Also explicit from the Committee's process is the amount of time given to various witnesses during hearings. In all, the Committee conducted 98 sessions of testimony during the 32 public hearings, taking in sum 80 hours and three minutes.[11] How they decided on (or ended up) allocating that time speaks, in some degree, to the access that individuals and organizations had, and it speaks to the value placed on that session by the Committee. Generally, the time allotted to witnesses was fairly uniform, with an average of 49 minutes given to each testimony session. To determine who was listened to most—in quantity anyway—I constructed lists of those allotted time one standard deviation or more above and below the mean. Those allotted the most

Table 3.2 Cross-tabulation of Methods of Analysis for Agreement Between Informants and the Committee

Name	Submission agreement score (≥+1SD)	Submissions and hearings cross-referenced (≥ median)	Explicit citations, by source (≥ mean)	Time allotment (≥+1SD)
Association of Heads of Independent Schools in Australia	■	■		
Association of Independent Schools of SA		■		
Australian Association of Social Workers	■	■	Submissions	■
Australian Council of State School Organisations	■	■		■
Australian Education Union	■	■	Submissions	
Australian Hearing			Witnesses	
Barresi*		■		
Bartlett*		■		
Bednall	■	■		
Board for Lutheran Schools		■		
Boys in Focus		■	Submissions	■
Browne		■	Submissions	
Buckingham	■	■		
Button	■	■		
Butz	■	■		
Canberra Grammar School	■	■		■
Catholic Secondary Principals' Association of WA	■	■		
Crews			Witnesses	

Continued

Table 3.2 Continued

Name	Submission agreement score (≥+1SD)	Submissions and hearings cross-referenced (≥ median)	Explicit citations, by source (≥ mean)	Time allotment (≥+1SD)
Dept. of Education and Community Services, ACT		■	Witnesses	
Dept. of Education, SA			Submissions	■
Dept. of Education, Tasmania				■
DETYA (a.k.a., DEST, DEET)	■	■	Exhibits, Submissions, References	■
Education Queensland		■	Submissions	■
Elanora Primary School, Queensland		■	Witnesses	
Endeavor Forum	■	■		
Festival of Lights, SA		■		
Fletcher (Men & Boys Program, Univ. of Newcastle)	■	■	Witnesses, Submissions, References (2)	
Fremantle Education Centre		■		
Griffith Public School, NSW		■		
Hill (Univ. of Melbourne)		■	Witnesses	■
Humphreys	■	■		
Hurstville Boys' High School	■	■		
Independent Education Union	■	■		
International Boys' Schools Coalition		■		
Ken and Katherine Rowe	■	■	Exhibits, Submissions	■

Continued

Table 3.2 Continued

Name	Submission agreement score (≥+1SD)	Submissions and hearings cross-referenced (≥ median)	Explicit citations, by source (≥ mean)	Time allotment (≥+1SD)
Kerry Davies	■	■		
Kormilda College, NT		■		■
Lewis		■		
Lillico		■	Witnesses	
Mabel Park State High School		■		
Ministry of the Premier & Cabinet, WA			Submissions	
Mullins		■	Submissions	
Nelson*		■		
New Town High School, Tasmania		■		
NSW Dept. of Education and Training			Submissions	■
NSW Secondary Principals Council	■	■		
NT Dept. of Employment, Education & Training		■	Witnesses, Exhibits	■
Palmerston High School, NT		■		
Queensland Catholic Education Commission	■	■		
Randwick Boys' High School	■			
Religious Education and Educational Services	■	■		
Roseville Public School, NSW				■

Continued

Table 3.2 Continued

Name	Submission agreement score (≥+1SD)	Submissions and hearings cross-referenced (≥ median)	Explicit citations, by source (≥ mean)	Time allotment (≥+1SD)
SA Association of State School Organisations	■	■		
SA Independent Schools Board	■	■		
SA Primary Principals Association		■		
Sawford*		■		
Skyvington		■		
Stanbridge	■	■		
Tallebudgera Beach School, Queensland		■	Witnesses, Exhibits	■
Tintern Schools		■		
Trent and Slade (Flinders Univ.)	■	■	Witnesses, Exhibits, References	
Trinity Grammar School, Victoria		■		
Victorian Government	■	■		
WA Council of State School Organisations		■		
Wade High School			Witnesses	
West (Univ. of Western Sydney)	■	■		
Woodridge State High School, Queensland		■		
Yates			Witnesses, Submissions	
Yenda Public School		■		

Note: Shaded rows are those witnesses that had three or more categories present.
■ = The witness meets the specified criteria.
* = The person listed was a member of the Committee.

time (see table 3.2) were generally state and Commonwealth education departments, a few schools, and academics the Committee liked. Those below the mean were largely student groups interviewed during school visits. These students were often taciturn and could hardly be expected to talk about many of the issues specified in the terms of reference.

Pulling the Analysis Together:
Who was Listened To

Looking across these multiple methods—the hearing analysis, the submission analysis, the cross-referencing of hearings and submissions, the explicit citations, and the time allotment—a clearer picture of the political contours of the report develops, for many overlaps exist on these lists.

Table 3.2 shows the cross-tabulation of the various methods described above for determining the relative influence of various actors on the Inquiry and its report. As one can see from the table, where those shown as influential by three or more of the four methods above are shaded, a small group of people and organizations had tremendous influence on the shape of the policy. Thirteen groups and individuals show such significant influence, all of which I detail below.

Not being included on this list, of course, does not mean that a witness or submitter had no influence. Doctors LePage and Murray from Australian Hearing, for example, didn't make the list, but they were extremely influential. Their testimony and exhibits were the sole reason for including a section in *BGIR* on boys' hearing and were a major component of recommendation five, suggesting hearing and vision screening for all children.[12] Other key influences are also missing, like Ian Lillico, a famed Australian popular-rhetorical writer on boys' education. Perhaps if Lillico had written a submission, he would have made this list. Nevertheless, some of the people and organizations included on this list make for a "who's who" of policy influence on boys' education and even other issues.

While the preceding analysis gives some indication of the *who* and *what* of influence on the Committee and its policy—a useful function for such quasi-quantitative measures—it does not give a particularly good picture of the *how* and *why*. For that, consider the following qualitative descriptions of these influential people and groups (listed alphabetically).

Australian Association of Social Workers

BGIR mainly relies on the Australian Association of Social Workers, a group of 6000 members, to argue for the rather progressive notion that single-parenthood isn't the cause of social problems in and of itself, but rather other problems tend to coincide with single parenting (*BGIR*, 54–55). The Committee was also shown a violence prevention program the Association organized. Relatively ignored from the Association's submission, though, were the stress on poverty and rural location as causing educational failure and the assertion that the focus should particularly be on disadvantaged boys.[13] Also overlooked was the Association's stress, both in the submission and in testimony (June 7, 2001), on making schools into community service centers, though the Committee did recommend some health screening (recommendation 5).

Australian Council of State School Organisations (ACSSO)

ACSSO is an umbrella organization providing support to parent groups and school councils across Australia as well as to its state-level chapters. The reportedly two million parents in the organization represent a key constituency to the Committee members. In *BGIR*, cited evidence from ACSSO is used to suggest several measures of post-school outcomes that currently are ignored in policymaking (*BGIR*, 36). Many of their recommendations, however, are also recommended in *BGIR*, including hearing and vision screening (recommendation 5), increasing teacher pay to attract more males (recommendation 18), and training teachers to address gendered learning styles (recommendation 2).[14] Overall, ACSSO's arguments aligned with the final report, though they emphasized more the influence of other social factors with gender and providing more social services through schools.

Australian Education Union (AEU)

The AEU is Australia's largest teachers union, with a membership approaching 155,000 and comprising teachers in schools, preschool centers, Technical and Further Education institutions, and other educational settings. While this is also an important constituency for Committee members, both as votes and for implementation of any policies, the small size of this group make it less powerful than other organizations. The AEU also has a longstanding reputation for progressive, (pro)feminist politics, a fact that made it a target in this remarkably antiunion Committee. Thus, in *BGIR* the AEU is primarily used as a foil for the Committee's positions. For example, *BGIR* paints the AEU as oppositional to achieving good outcomes for all students.

Most of the teachers who have contributed to the inquiry, whether male or female, have been eager to address boys' education issues as part of their commitment to achieve the best outcomes for all their students. However, it is difficult to avoid the impression that some gender equity units in education departments and education unions, generally, have been reluctant to openly confront boys' under-achievement and disengagement as an issue, perhaps for fear of undermining ongoing support for strategies for girls. (61)

On only one occasion—apart from quoting a statistic—are the Union's views presented positively: discussing the negative views society holds of teachers. The AEU's arguments for paying teachers more and reducing class sizes are supported in *BGIR* (recommendations 18 and 13, respectively), but largely these are decontextualized from the Union's urging against simplistic solutions to boys' education concerns, which the AEU largely considers "backlash."[15]

Boys in Focus
Boys in Focus is a group of educators from various service backgrounds that have worked with troubled boys in New South Wales since 1992.[16] Their successes have been chronicled in an article in the University of Newcastle's popular-rhetorical, practitioner oriented newsletter, the *Boys in Schools Bulletin* (volume 2, number 3), for which they were cited in other submissions, including by popular-rhetorical consultant Richard Fletcher, discussed below.[17] In *BGIR* the group is cited on negative images of males in the media (90), gang involvement when family and school support breaks down (130), and the importance of strong relationships between teachers—especially males—and students (153, 161). More importantly, an entire section of *BGIR* is devoted to Boys in Focus (132–4). In it, the Committee bemoans the lack of structural support for finding, training, and supporting people to run programs like Boys in Focus without overburdening a few committed teachers. The Committee didn't, however, make recommendations that would accomplish such support.

Canberra Grammar School
Canberra Grammar School is an elite private, boys-only school located near Parliament House in Canberra. Its alumni include Larry Anthony, National Party member of Parliament; former prime minister Gough Whitlam; media mogul and, when alive, the richest man in Australia, Kerry Packer; and likely many children of current members of Parliament. Certainly its proximity didn't hurt its influence with the Committee, for the Committee was able to easily visit the school, only a short ride from

their offices. The school's only direct influence on *BGIR* was a quotation supporting the Committee's notion that relationships between students and teachers are key to boys' educational success (134). Nevertheless, Canberra Grammar's arguments closely matched the Committee's concern for relationships, the need for boy-specific pedagogies, and the need to have boys expand their skills and attitudes for the workplace of the future.[18] They were, additionally, the only submitter to suggest that the states not decrease their contributions if given extra funds for boys' education (*BGIR's* recommendation 24). Their suggestion, though, that schools encourage "new forms of masculinity... or at least encouraging the acceptance of forms which in the past were considered less than to be a 'real man'" was ignored.[19]

Department of Education, Training and
Youth Affairs (DETYA; later DEST)
The Commonwealth ministry of education, at the start of the Inquiry called DETYA, is naturally a major player in any educational policy endeavor. The bureaucrats, accountants, lawyers, and policy crafters in this department form the key national-level organization for numerous programs, initiatives, and policies. They also collect educational statistics and are the contracting agent for most grants and research projects. To not have DETYA as a key influence would be almost unthinkable in the Australian educational policy ecology. Largely, though, their influence on the report-making process in *BGIR* was limited to argumentless statistics and research findings.[20] This fits well the Department's reputation, historically, of being a resource of mainly "objective" data. Despite any lack of rhetorical flourish, the influence of DETYA's testimony at the first hearing made a huge impression on key Committee members, particularly one single line: "The difference between boys' and girls' average Tertiary Entrance Score, the NSW Year 12 aggregate, increased from 0.6 marks in 1981 to 19.4 marks in 1996."[21] This statistic stuck in the minds of Committee members, largely as proof of a boy crisis, for, if scores had actually *declined* over time, boys' disadvantages in literacy could not be explained by girls getting better comparatively; instead, something would have made the boys get worse. For Mr. Sawford, especially, this was a grave and continuing concern, for he brought it up 21 separate times in various hearings. For example, he recalls:

> ...at the first public hearing when the Commonwealth Department of Education, Training and Youth Affairs presented evidence to us, one of the comments that stuck in all of our heads—and they were referring to

Victoria, Queensland and New South Wales—was that if you compared the attainment levels of boys and girls in 1980 with what they are now, there was a variation. In 1980 the variation was less than one per cent, which you would expect because girls and boys are intrinsically equally as intelligent, so why wouldn't they be doing the same. Now the variation is up to 20 percentage points.[22]

Note that Sawford has, over the course of time, blown the statistic out of proportion. This result is only pertinent to New South Wales, not to the other states he mentions, and this test is a limited measure of boys' achievement. It gauges only 12th grade boys in a single state. Also, the DETYA submission goes on to explain that assessment methods may have played a part in this apparent decline, a point that Sawford takes up in many hearings, but the DETYA submission, if one reads further, also suggests that masculinity issues may have played a part, for "Boys are more likely to group in a narrower range of high pay off and/or traditional subjects with a higher risk of poor performance."[23] This latter conclusion, one that put to question the current norms of masculinity, the Committee never took up. This pattern of systematically avoiding any challenge to masculinity's hegemonic maxims in favor of pedagogical and assessment reforms occurs throughout the report.

Education Queensland (EQ)
Education Queensland was the name for the Queensland ministry of education during the Inquiry. EQ's boy-specific programming and policymaking had already, before the Inquiry, gone arguably farther toward separate boys' policy and male teacher policy than any other state. The EQ Boys' Education Reference Group authored the EQ's submission.[24] Other states had no such group. EQ also had a *Boys' Education Strategy*, which no other state had yet created. EQ was also among the first to have a taskforce devoted to improving the recruitment and retention of male teachers. Add to this the innovative curriculum and assessment plan, New Basics and its "Rich Tasks," which was being pilot tested at the time, and Queensland became an appealing case for the Committee. *BGIR* lauded the New Basics framework as having "the potential to significantly enrich the learning experiences of all children as well as improve boys' interest and engagement" (74). Overlooked or downplayed, though, was EQ's appeal to focus on other issues, like poor outcomes for Indigenous students and high unemployment in many areas. EQ also stressed that they were moving away from a "target group agenda" toward "intersecting factors" that combine with gender to produce inequalities.[25] The Committee, however, ultimately framed boys as a target group for policy

intervention, and the resulting initiatives reflect that target orientation (see my chapter 4).

Richard Fletcher
Richard Fletcher is an instructor at the University of Newcastle and a director of its Men and Boys Program, which runs workshops on male mental and physical health topics. He is also a popular-rhetorical author and consultant who typifies the "what about the boys?" strand of boys' education discourse.[26] His popularity around Australia probably explains why the Committee gave an entire hearing to him. During that hearing, the Committee was effusive with praise for his producing research that's "different" and a "substantive addition" to the Inquiry.[27] Perhaps his biggest contributions were criticisms of the *Gender Equity Framework* (*GEF*) and his insistence on recasting it to include boys' needs—criticisms that made their way almost verbatim into *BGIR*.[28] He was also explicitly cited in *BGIR* for strategies that support boys' learning, work on choice theory in discipline, encouraging fathers to participate in their children's educations, and providing male role models who value learning. So aligned was his testimony and submission to the Committee's and the Government's take on boys' education that Fletcher and his colleagues were awarded federal money for their Boys to Fine Men conferences and the recasting of the *GEF* along the lines of his criticisms (detailed in chapter 4).

Peter Hill
Peter Hill's reputation obviously preceded him, as well, for even without putting in a submission to the Inquiry, he was invited for solo testimony.[29] Hill was, when he testified, a professor and director of the Centre for Applied Educational Research at the University of Melbourne. His expertise in assessment was one reason the Committee wanted his testimony, for they had concern that assessment systems were disadvantaging boys. *BGIR* cites Hill on many issues, though, including the need for teacher reflection opportunities, the effectiveness of Reading Recovery, and the need for early literacy intervention. Indeed, for the latter, Hill was one of few suggesting that each school have a literacy coordinator, a recommendation the Committee made (recommendation 10). The Committee didn't agree with him on every point, though. Sawford, for example, took umbrage with Hill's notion that boys have a "developmental lag."[30] Perhaps more importantly, the Committee ultimately disagreed with Hill's stance that a balanced approach to literacy is better than a phonics focus. In fact, the Committee's recommendations focused almost exclusively on installing

phonics instruction in schools, programs, and teacher preservice and professional development (recommendations 7, 8, and 9).

Northern Territory Department of Employment, Education and Training (NTDEET)
Like Education Queensland, NTDEET had some influence on the report, though mostly limited to Indigenous issues. The Northern Territory (NT) has a large Indigenous population (only Queensland's is larger), and NT often gets associated with Aboriginal issues. In some ways this reflects the NT's in-depth, focused effort on Indigenous education.[31] In *BGIR,* only three citations of the twelve from NTDEET were *not* concerned with Indigenous issues, and those regard differences in assessment and course selection statistics. Only two other states were cited in the section on Indigenous education (*BGIR,* 31–36). As I explain later, despite NTDEET's influence on Indigenous education within the Inquiry, the Committee still made no specific recommendations on Indigenous issues.

Ken and Katherine Rowe
Ken Rowe is a principal research fellow for the Australian Centre for Educational Research, a premier Australian research body for education. His wife, Dr. Katherine Rowe, is a pediatrician at the Royal Children's Hospital in Melbourne. The two testified to the Committee together and cowrote their submissions, though most of this information originates in Ken's work.[32] He, particularly, has been visible in national education debates and has been a featured speaker at conferences, including the Boys to Fine Men conferences described in chapter 4. Rowe is charismatic, but even more he fits well the quantitative orientation that seemed to most capture the Committee's attention (discussed below). Indeed, Minister Nelson was so impressed with Dr. Rowe's research and beliefs that he appointed Rowe as chair of the high-profile National Inquiry into the Teaching of Literacy 2004, a taskforce that ultimately recommended the same literacy approach *BGIR* did: heavy phonics emphasis.[33] Phonics, though, wasn't Rowe's only influence, for the Committee cited, among others, his complaints about assessments' increasing literacy demands, his strategies that "support the learning needs of boys" (*BGIR,* 79), the central role of teachers rather than socioeconomics (85), the superiority of single-sex schooling on student achievement (86), and the key role of professional development in making reform work (implicating four of *BGIR's* recommendations). Perhaps most influential, though, was the Drs. Rowe's testimony about auditory processing—the ability of the brain to handle

and interpret what is heard—and a Victorian professional development project on accommodating differences in auditory processing. Part of *BGIR's* recommendation five suggests this program be implemented across Australia. The Committee had few, if any, substantive disagreements with the Rowes.

Tallebudgera Beach School, Queensland
The alternative school program for at-risk youth, called 3R, at Tallebudgera Beach School on Queensland's Gold Coast, invited the Committee to their school, largely to secure funding for expanding their program.[34] While they got no funding, they did exert important influence on the report. The 3R program was featured as a model alternative program (*BGIR*, 149–51) and was one basis for recommendation 15, that research be funded to compare the effectiveness of various school structures, including alternative schooling. The witnesses also shared the Committee's concerns with reducing class sizes (recommendation 13), focusing on teacher–student relationships (e.g., *BGIR*, xx), and making the curriculum relevant to students (*BGIR*, chap. 4). Perhaps the Committee's somewhat tentative endorsement was due to some members' concerns about such programs being "social engineering" and about its exportability.[35]

Faith Trent and Malcolm Slade (Flinders University)
The Committee was impressed with Trent and Slade's DETYA-funded work interviewing boys about their attitudes toward school.[36] Though the study was largely qualitative (an issue of concern for Mr. Sawford), the sample size—1800 students, in focus groups of ten—seemed to sway the Committee to its importance.[37] The study paints a dire picture of boys' attitudes, for the boys reported that teachers don't listen to them and curriculum and pedagogy are irrelevant and even insulting. The Committee largely used this research to support their case that curriculum might cause boys' "disaffection" with schooling (*BGIR*, 143–4) and that boys' lives out of school conflict with schools' expectations (*BGIR*, 66). The Committee, though, neglected Trent's and Slade's testimony that girls believe the same things but are not so vocal about it.[38] In *BGIR*, the Committee presented the findings as relevant to boys only—as proof that schools work against boys—when actually the broader implications of Trent and Slade's work should have mitigated that view.

Analysis
If there were doubts about the veracity of the list above being representative of major influences, it is telling that many of the informants in the

preceding set, particularly the individuals, were pointed out as important by Committee members themselves in the hearings. Mr. Sawford, for example, tells Richard Fletcher,

> You mentioned, from academia, Ken Rowe and by inference you mentioned Faith Trent and Malcolm Slade. We have had Professor Peter Hill to give evidence. They put forward a view which I think to most members of the committee, whatever their politics, comes across as a very commonsense sort of approach.[39]

All of those witnesses that Mr. Sawford speaks positively of, including the person spoken to, are on the list of those most influential, those with the loudest voices in the rewriting of gender policy for Australia.

How can one understand *why* certain informants were listened to over or more than others? Understanding this helps give shape to the political contours of the Committee and its process. First, several *kinds* of people and organizations were validated through the Inquiry's process:

1. The organizations are mostly long-established groups with particular responsibilities for education and service provision.
2. Several state, territory, and Commonwealth education departments made the list, though their influence tended to be mainly descriptive of state contexts.
3. Individuals with influence tended to be academics.
4. Individual academics tended to be researchers doing large-scale, often quantitative, research.
5. Most on the list were interconnected professionally (for example, Slade and Trent did their work for DETYA, Hill and Ken Rowe worked together, Fletcher's journal published the work of Boys in Focus, etc.).
6. All individuals were white and middle class and, because they didn't explicitly say otherwise, their research implicitly represented the interests of those groups.

I might also add that most on this list represent a politically conservative slant. The one truly (pro)feminist organization, the Australian Education Union, was again influential only as a target for the Committee's largely antiunion sentiments. True, the Committee avoided the most extreme antifeminist arguments presented to them, but most arguments were decidedly conservative.[40]

In fact, these conservative leanings were evident from the start of the Inquiry. Consider the following statements by members during only the fourth hearing (October 26, 2000), a mere three weeks into the hearings. Mr. Sawford said to members of Tintern schools who were testifying that

> You [Sylvia Walton, a principal] have not mentioned gender, you have not mentioned socioeconomic status, you have not mentioned homophobia, you have not mentioned a whole range of other things that have been put to us in terms of determinants. What you have put to us is, in fact, balanced.[41]

While clearly rhetorically framing an apolitical stance as "balanced," Sawford ironically forwards a conservative political stance—that social inequalities are beside the point—as preferable. Moments later, a Coalition member, Mr. Barresi, gives similar indication of the arguments he intended to validate:

> ...a number of witnesses, particularly those I have to say, *regrettably,* who are in the bureaucracy or in decision making positions who dispute what you are saying about the verbal reasoning that is going into curriculum or even the importance of male teachers versus female teachers and the fact that there is even a difference between girls and boys in curriculum. (emphasis added)[42]

Already, by the fourth hearing, Sawford and Barresi clearly outline the arguments they were willing to accept, and by no surprise, the final report, two years later, reflects these largely a priori decisions. One might be forgiven, in fact, for underestimating how much the public hearings were opportunities for Committee members to assert their *own* views. I came to my analysis of the process—somewhat naively—anticipating that Committee members were primarily listening. Yet they often approached contrary views antagonistically, made unsolicited declarations of their own positions, and harped on points upon which they had already decided.[43]

Who Wasn't Heard: The (Pro)Feminist Resistance

On the opposite end of the Committee's traditional beliefs about boys and their educational needs, the views that they approached antagonistically, as I just said, were those of progressive and (pro)feminist scholars, educators, and social service providers. At the conclusion of the fifth chapter in *BGIR* the committee writes, "There are several reasons for boys' poorer performance in literacy compared to girls. These reasons are not fallacy or folklore" (127). Here, in barely coded language, lies the Committee's reaction to what they thought the (pro)feminist resistance to boys' education reforms were arguing. Why this characterization, that the (pro)feminist resistance believed boys' difficulties to be "fallacy" or "folklore"? How could the Committee so thoroughly misinterpret and dismiss the arguments that (pro)feminists had made throughout the policy process?

Examining the considerable resistance to the report-policy is crucial to fully understand the political and social dynamics at work in Australia and abroad. Though neither side of the boys issue has annihilated the other, one faction is flourishing at present while the other struggles to have its voice heard. Make no mistake: the conservative, "what about the boys?" factions in Australia and elsewhere are "winning" in the political and policy sphere. This victory, however, is not complete, not totalizing, and not irreversible.

In trying in this section to comprehend why this resistance has largely been unsuccessful in preventing boys' issues from dominating the equity agenda, I have some serious critiques of the (pro)feminist resistance. Two caveats are in order, though. First, I want to avoid giving the impression that the "(pro)feminist resistance" is a coherent or organized group. It is, rather, more a circle of like-minded scholars, bureaucrats, educators, and community workers typically with a degree of connection to one another and a penchant for collaborating, but it lacks structure and coordination. This is generally true of the "other side" of the debate, too. Second, I want to make clear my intentions in writing this section, particularly the critiques of the progressive, (pro)feminist resistance. Michael Apple, great scholar of the politics of education, recounts a joke that his grandfather told about political leftists, saying that, when they line up for a firing squad, they line up in a circle, guns pointed inward.[44] While political infighting is common in any ideological camp, the joke's kernel of truth is instructive for progressives. I am loathe to wound friends and colleagues in the (pro)feminist camp, those who supported my work and with whom I agree 98 percent of the time. My intention is definitively *not* to blame them for what has happened, for their fight has been strong, honorable, and often effective. Still, they were competing within an ecology hostile to their interests and stacked against them in political power, money, and communicative dominance; the situation is quite similar for (pro)feminists in the United States. My real task is to understand why things happened as they did so that battles in the years ahead might be more successful. Furthermore, I have the benefit of hindsight, being able to view cold transcripts rather than having to be articulate in heated exchanges. Quite often, contexts alter and become toxic more quickly than those in it can realize and cope. This book, though, isn't a postmortem; it is an analysis for the means of better evolving to meet the challenges of a changing—and currently hostile—ecology.

(Pro)Feminists have largely resisted the moves toward boys' education policy and practice from a belief that the turn to boys is unwarranted, overblown, or dangerous to girls' gains. In Australia, this resistance goes back at least to the defense of girls' education policies of the 1980s and 1990s (see chapter 2). Debra Hayes, in examining the changing maps of gender equity discourses, suggests that resistance worldwide has taken three forms: rejection accommodation, and transformation.[45] Strategies of *rejection* have typically entailed rearticulating that girls still have problems and deserve research and practitioner attention. Hayes points out that such rejection moves, which are easily ignored or accommodated by boys' education discourses, can, because they're counter to the now dominant boys-focused narrative, "appear somewhat nonsensical, irrational even."[46] Over time, furthermore, such strategies become increasingly ineffective. As Hayes explains,

> Subsequent [rejections] are removed more efficiently as new tactics for dealing with oppositional statements are incorporated and co-opted within the dominant discourse. As new discursive regimes [around boys' education] consolidate through the emergence of groups of experts, the establishment of centres of knowledge and the legitimation of policy support, then rejection strategies appear increasingly feeble and those who articulate them are increasingly vulnerable to marginalization.[47]

Much of this has clearly happened to those maintaining a purely rejection-oriented strategy, for the "commonsense" status of boys' educational needs, so prevalent in the media and practitioner professional learning, those who only reject boys' education have increasingly come to look "out of touch" with the perceived "reality" of boys' disadvantage.

The second strategy, *accommodation,* is employed in an "already altered terrain."[48] Here oppositional forces attempt to redirect the dominant boys discourses, often attempting to salvage particular components of the girls discourses and to moderate the most regressive components of policies or programs. The "Which boys? Which girls?" questions grow out of such a strategy. As Hayes rightly points out, though, such accommodation moves have already "capitulated some ground" to boys discourses, which makes it *more likely* that the remnants of girls' issues will be pushed aside or reshaped to fit the dominant discourse.[49] Hayes also suggests that coalitions and alliances, with men or with homosexual movements, can also be a form of accommodation, though these movements tend to be marginal and temporary.

Hayes presents the final strategy, *transformation,* as more of a speculative rather than widely practiced strategy in the (pro)feminist movement. In transformation, those resistant to the boys' education discourse locate "intrinsic defects" and "points of weakness and fragility" within the dominant discourse and exploit those to show its dangers and to show that other possibilities—progressive ones—exist.[50] In this strategy, then, the weaknesses of the new movement are transformed rather than rejected outright. Again, though, such a strategy has largely yet to be utilized.

Despite the successes of those mounting the boys' education policy drive, boys advocates have shown they still need to *constantly* establish their own legitimacy. Politicians pushing this discourse had to launch research projects to back up their conclusions, careful to invite those that wouldn't be seen as partisan.[51] The striving for legitimacy is also evident in the co-opting of discourses from (pro)feminists, especially the question of "which boys?" and the need to not hurt girls' gains. Importantly, though, these are arguments that pose no profound threat to the initial interest in boys. Still, acknowledging the entry of the "which boys?" and "don't hurt girls" discourses is crucial, not only to demonstrate the political maneuvering of boys' advocates, but also because this co-opting shows that concerted resistance by (pro)feminists *does* make a difference and that boys' education has progressive potential.

Still, much of the progressive promise of boys' education has largely been unrealized. The (pro)feminist resistance has indeed had some successes, but it has also made strategic mistakes that have either helped the conservative boys' discourses grow or limited progressive discourses.

Firstly, and out of their control, (pro)feminists were hampered by Labor politicians' *contributions* rather than resistance to the policy. As mentioned earlier, one of the most vocal crafters of the policy-report and its conservative realization was the Labor deputy chair, Rod Sawford. His lambasting of unions and academics that resisted the policy clearly showed his conservative leanings. Furthermore, the most vocal skeptic of boy crisis discourses, Julia Gillard, left the Committee and thus could not help steer it to more progressive ends. Labor's position on *BGIR*—it was after all a unanimous report—demonstrates a turn to the right on gender equity, or even an abandonment of feminist positions. This has made (pro)feminist resistance that much more difficult.

Part of the difficulty for allying with the Committee's Labor members was the shifting common sense around boys' education to which politicians, just as the public, were constantly exposed. In many instances, the arguments that (pro)feminist scholars and educators presented existed outside this "new" common sense, and (pro)feminists either didn't successfully explain their concepts or they failed to correct some key "mishearings" by the Committee. One such concept is the social construction of gender. Three times in the March 2001 hearing in Adelaide, for example, Mr. Sawford asked witnesses to tell him the *quantitative* evidence underpinning social construction. He also suggested in a hearing the next month that the whole concept of social construction was questionable because teachers and principals didn't themselves use such terms.[52] This latter view seemed to prevail in *BGIR* itself, for the only mention of social construction of gender is used to criticize unions for suggesting "too much emphasis on gender theory" (*BGIR*, 160). Sawford has clearly misunderstood the application of the gender construction concept, and, rather than differentiating this concept as a theoretical, not empirical, understanding of how gender works or simply contrasting it with biological gender, those who answered Sawford on gender construction responded largely to his demands for quantitative evidence. So, instead of defending social construction—also under attack from other conservative "what about the boys?" advocates (see chapter 4)—those witnesses ended up rather defending the value of qualitative research.

Another key example of a mishearing that went unclarified was the role of socioeconomic status, race, and other subjectivities in influencing boys' outcomes. A recurring question that seemed to plague the Committee was whether socioeconomic status *determines* school outcomes or whether quality teaching can overcome it. This hits right at the heart of the "which boys, which girls?" argument put forth by (pro) feminists, and it shows that the Committee is indeed concerned about such issues. The Committee members seem implicitly, though, to reject the notion given by (pro)feminist and progressive witnesses that socioeconomic status is more of a problem than gender. The Committee seemed to read these contentions as a deficit model, one that blames the disadvantaged rather than structured inequality, as if an excuse for low expectations. Dr. Nelson said:

> ...[A] number of people have basically said to us, "If the kids are poor, they're coming to school underfed," and all that sort of stuff. You basically take a defeatist attitude—that is what they have implied—whereas others have said,

"Where the rubber hits the road, if you've got enthusiastic, committed, well trained, professionally developed teachers, then you'll get a better result."[53]

Sawford echoes this:

> Something has gone wrong somewhere. It seems that when we talk to people, in public education mainly, there is almost a sense of denial first. Then they go back to what Mike [Owner; a witness] was saying this morning in terms of deficit models which I do not think help anybody and then they deal with the periphery. We have dealt with gender construction, gender equity, homophobia and goodness knows what else. I did not think [these] had a great deal to do with [boys' education]. I am serious.[54]

Sawford clearly reads sociological inequity arguments as irrelevant, or at least "peripheral." Often, however, (pro)feminists seemingly misrecognized this criticism. They put forward arguments that one could not simply look at all boys as being disadvantaged, but the Committee clearly viewed arguments about targeted need as a deficit view, and few (pro)feminists had an answer for this.

The (pro)feminist resistance also ran into difficulties in winning support because they rarely offered help solving the gritty materialities of boys' issues. Much of the boys' education turn, as I said in chapter 1 and elsewhere, has been pushed by parental and teacher concern for the boys they work and live with daily.[55] They want to know what to do when boys are troubled, disengaged, badly behaved, or taking risks with their lives and others' lives. Consider for instance the concerns of one parent:

> My son commenced his education at a state school at the age of 5 years. Even at that age he displayed signs of not being able to keep up with the rest of the class. At that time I was advised by teachers at his school that he was a boy, they are slower developing and not to worry. At no time was he or any of the other boys experiencing difficulties given any extra help or was any advice given to parents in relation to what can be done out of school hours to help. [As written][56]

To look for concrete answers to such practical dilemmas is reasonable. Indeed, this was one of the terms of reference of the Inquiry. Concrete strategies for practice have been somewhat rare from the (pro)feminist resistance, though. Exceptions exist, of course, but a main tactic of the (pro)feminist resistance has been to provide theoretical reasoning for why boys are not a problem, a notion contrary to the lived experience of many educators, parents, and social workers.[57] Practitioners are left to look for help from conservative popular-rhetorical texts with a practitioner focus.[58] This, again, is how such texts become de facto policy.[59]

Also, (pro)feminists missed a chance to advocate for a "balanced" approach that would support the justifiable academic needs of boys while attacking a broad spectrum of social ills. In other words, opportunities were missed to put topics like antisexism, antiviolence, and antihomophobia within a social program of sex education, anti-bullying, and self-esteem. By marking themselves early as resistant to boys' education concerns, though, some (pro)feminists may have taken themselves out of the running to administer and shape programs being funded for boys, such as the Boys' Education Lighthouse Schools program and the *GEF* recasting.

(Pro)Feminists might also have seemed less "in denial" by doing more to recognize and build on the contributions of boys' education advocates and the progressive potentials of those advocates' programs. Instead, (pro)feminists largely ignored these. They could have, for instance, built on and supported: (a) pedagogical advances that are good for all students, such as building good relationships between teachers and students, utilizing learning styles that are not always valued in schools, and encouraging active involvement; (b) bullying-prevention programs and behavioral interventions that help all students; (c) opportunities to recognize that boys have gender and explore it through critical literacy; or (d) proposed changes to labor conditions of teaching like pay increases, which, though geared to attracting more men to the profession, would also benefit women.

The final strategic mistake regards (pro)feminist arguments that ask the public and teachers to suspend their "common sense" about gender equity, a common sense that (pro)feminists themselves helped create. To illustrate, consider what Mr. Barresi, a Coalition member of the Committee, asked of two administrators:

> One of the two areas which I would like to get your views on that has been disputed is that it does not really matter that the boys are underperforming the girls in school because at the end of the day in employment that evens itself out. They will get a job....
> The second one is that it does not really matter that 70 per cent of our teachers are female because the hierarchy within the school and the education system is male dominated anyway....[60]

These are common arguments, ones that are part of what I call the *counterintuitive 'big ask.'* They're "counterintuitive" because they contradict "commonsense" notions that the public uses in determining equity. It is a "big ask"—in the Australian vernacular, a request that's considered difficult to meet—because such arguments require overlooking other key elements of schooling. The first that Barresi mentions asks the listener to overlook the negative experiences of school that some boys face—like harassment, bullying, ineffective teaching, and lesser skill acquisition—just because they're likely to do well in the workforce. This argument denies that *some* boys are suffering in their *current* context, thus abrogating educators' responsibility to ensure that schools are safe and effective, even denying some boys their right to basic skills. The second argument, that it does not matter that there are fewer male teachers because there are more male administrators, also suggests that listeners overlook the fact that many males don't feel comfortable choosing teaching or staying on in teaching and that there could possibly be benefits for students in having positive male teachers.[61] In both cases, then, the listener is presented with a counterintuitive big ask: to overlook what, on the surface, appears to be basic educational rights and equities.

There are political ramifications of making counterintuitive big asks in terms of guiding policy and drawing in allies, though. To illustrate, Richard Fletcher, discussed above, testified,

> ...my main concern is the way labour market outcomes are the only things we talk about. You look at the boys in high school and the girls in high school and say, "When they leave here, the boys are going to get more apprenticeships in the metal industry. The boys are going to get higher paid jobs when they are 19, on average, so we should address that." True. We also know that for every girl that dies in high school, over 15 [years old], three boys will die. That is a horrific statistic for Australia.... We should say, "What about that?" That is a reasonable outcome to look at as well, isn't it?... [W]hen we go to parent meetings no parent says, "Well, I don't care much about his health, I just want him to be able to use a computer." Parents do not say that. Of course they are concerned about their kids' lives. They want them to live good lives. We have been far too narrow in the discussion.[62]

Clearly, conservative boys' education advocates have seized upon this counterintuitive "big ask" and the material concerns of parents and teachers to effectively counter girls' needs, and they've used those understandings to make progressive scholars look even more "out of touch" and "in denial."

The (pro)feminist resistance, as I have suggested, has had some successes in holding off the worst tendencies of the boys' education movement in Australia, but it has also been put in strategically difficult positions and has made some strategic mistakes. Much has been lost by, and taken from, these (pro)feminists. I don't believe that all is lost, though. In the Coda,

I present reasons to hope that the worst, most regressive tendencies of the boys' education movement will be eradicated.

Constructing Authoritative Voices and the Cycle of Legitimacy

The selection of evidence that conforms to an a priori belief, whether about parenting or member's political stances, isn't unusual in policymaking. This result (not really a practice) raises some serious concerns for the larger policy context, however. It sets up what I term a "cycle of legitimacy" that can exclude and ultimately wither key understandings and theoretical positions.

To define it concisely, a cycle of legitimacy occurs when a group that seeks or has authority relies for its own legitimacy on arguments that it first validates. They legitimize the research that legitimizes their report. Though seemingly tautological, my argument rests on the realization that many of the positions presented to the Committee were highly ideological and contested. Even considering many of the "facts" presented (boys' literacy scores are below girls', boys are more often disciplined, and so on), there are questions about *which* facts ought to be used in constructing a case. The Committee surely understands this, for it is the basis for part of their first recommendation, that broader indicators be used to measure which gender is "disadvantaged." The legitimacy of *BGIR*'s case about boys, though, must come from some authoritative evidence; otherwise, it would lack legitimacy and influence with stakeholders. If the evidence remains contested, though, how can they assert the legitimacy of their position? *BGIR must assert who speaks authoritatively.* It must grant legitimacy to only those informants that conform to its positions.[63]

Just as "right" boys are defined and made through this discursive process, the Committee must also allocate legitimacy to what it considers "good research," "good programs," and "good teachers." Take for example *BGIR*'s description of two studies on hearing. The report says:

> The Victorian research, reported by Drs Ken and Katherine Rowe, examined the ability of children (with apparently normal hearing) to listen to and recall information. The Australian Hearing data, reported by Drs LePage and Murray, deals with the amount of auditory information that the brain actually receives. The two reports are not concerned with precisely the same thing but *both are credible, scientific and objective evidence* that, on average, boys' capacity or ability to receive and process auditory information is less than that of girls. (105; emphasis mine)

What is at issue isn't the quality of these studies. Rather, the issue is that a political body not qualified to determine research's "scientific" value or "objectivity" has done so anyway. This is a *political* determination, not a scientific one. The fact that their determination should be presented as if it is somehow apolitical and authoritative undermines credibility and integrity. That it should be backed up by coercive policy means and funding on a large scale proves worrisome. They've created a cycle of legitimacy to support their own beliefs about research and their own contentions, many of which, as I showed above, they believed when they began.

Furthermore, the cycle of legitimacy continues after the policy writing is complete. Those legitimated sources—the academics and organizations *BGIR* has validated—have incentive to in turn legitimate *BGIR* for their own legitimacy. Those who have been delegitimated, like the AEU, have incentive to fight it. Also, those who want the resources that come with the policy initiative have incentive to take up, or at least speak using, the legitimated language of the policy, particularly when such language is a prerequisite for being rewarded those resources. This is just what happened for schools, as I explain in chapter 4.

The Power of "Parental Identification"

One major filter through which Committee members and their informants viewed and understood boys' education debates and decided legitimacy for particular people and discourses was parental experience (and, similarly, teaching experience). Though parenting is shot through with political and ideological inflections, and though using parental experience in creating policy is a political and ideological act, the witnesses and Committee members expressed grittily human and material concerns that transcend political identification: the desire to have one's children educated well and to be afforded safety, comforts, happiness, and opportunities. Still, using parenting as a frame of reference for policymaking has particular implications for *whose* parenting is validated, including issues of race, class, ethnicity, and religious tradition.

Take, as one example of parental identification as an epistemological filter, a story Mr. Bartlett, the Committee chair, told me:

> ...I mean even [*inaudible*] our own son almost *quoted* what had been given to us by a number of other witnesses about literature and English subjects being asked—expressing his frustration being asked to express his feelings about particular novels. He came home one night and said, "What do you mean what do I feel about this? I don't feel anything. It's a story. You know,

I can tell you what happened. I can tell you who the hero was. I can tell you how it ended. I can tell you the plot. But don't ask me what I feel about it." Because I think for boys—it's a generalization—but I think for boys there's not that engagement with characters in a novel as there perhaps is for girls and they'd [boys would] operate more on the cognitive level rather than that sort of emotional and that personal level. So to be assessing boys on how well they describe their feelings toward particular characters in a novel, I think, is a bit tough for them and disadvantages them relative to girls.

While I find it ironic that Bartlett validates expressing emotion when it is *frustration* about schooling but not *connection* to fictional characters, what is key here is that parenting becomes an interpretive frame for understanding the information he gleaned from the Inquiry. He traces a line from witnesses' arguments to parenting experience to a policy conclusion about assessment. That policy conclusion he clearly has forged based partly on personal experience with his own son. This experience, though, is integrally shaped by particular cultural, racial, and religious understandings of appropriate emotionality and parental roles. To unpack this, though Bartlett obviously shares a universal concern that his son be fairly educated and assessed, his parenting frame is situated in a white, Anglo-Australian ethic of masculinity as emotional disconnection and suppression, so instead of encouraging connection in his son, assessments in his view should be changed to privilege the masculine approach of his cultural group. Most people understandably use their own experiences as frames of reference to interpret the world. Of concern, though, Bartlett has the power to make policy on the assessment of an entire nation based on such culturally and religiously inflected understandings. That is just what happened in *Boys: Getting It Right*.

Assessing the Report-Policy

As Michael Apple asserts is true of nearly any ideological or political position, there are elements of both good and bad sense in *BGIR*.[64] Exploring these positives and negatives gives a sense of the larger dynamics the policy participates in and its tenor and impacts.

Elements of Good Sense

While the "what about the boys?" debate has involved backlash against and forgetting of the gains made for girls, we should note that this discourse has influenced policy and research in *productive* ways too. It has, for instance, promoted interest in masculinity studies within educational research, which is key to understanding *some* of girls' oppressions and boys' difficulties. It has also shown many that boys *have* a gender, a notion that opens the door for potential male allies.[65] Including boys in gender discussions has also led to a sharpening of fundamental terms and concepts, such as "equity," "achievement," and "outcomes." Putting a generally advantaged group into these categories has led in some instances to a rethinking of the terms themselves. The "what about the boys?" debate has brought attention to social problems for boys that were, frankly, misunderstood as youth issues or popular culture issues rather than as issues of destructive masculinities, including suicide, criminality, violence, bullying, and drunk driving.

Similar to these larger debates, the Inquiry itself had positive outcomes that advance progressive initiatives. First, the Inquiry allowed the Committee to explore various pressing issues in education that may not have gotten attention otherwise, such as pedagogy, assessment, teacher pay and recruiting, health, and disadvantages other than gender. They addressed issues that progressive educators have struggled for: better remuneration for teachers, lower class sizes, and even challenging the limited roles available to men in popular media.

It bears mentioning, too, that conspiracy theories should be scrupulously avoided. This Committee acted responsibly in numerous ways, ways in which one would hope their government works. They opened the process to democratic participation, responded to numerous constituents by pursuing this topic, tried to elicit empirical evidence as well as stories of practitioners to back up claims, released a "unanimous" report, and made explicit attempts (with varying degrees of success) to strike a careful "balance" in political positioning.

Several features of the report, I believe, show the successful fruits of the attempt to be "balanced" and show, as I explored earlier, the successes of the (pro)feminist resistance to the "what about the boys?" movements. Firstly, the Committee uses nuance that the general debate on boys does not. Take for instance this critique of the debates on boys from *BGIR*:

> Unfortunately, much of the public debate about the educational underachievement and disengagement of boys has not gone beyond the simplistic idea that educational authorities should reverse an apparent imbalance in educational provision in favour of girls. Boys' needs are more complex than that implies and, in any event, girls' needs have not been universally met. (42)

The more simplistic arguments presented to the Committee eschewed such nuance, but the Committee in some ways successfully countered them.

Secondly, though they critique questioning "which boys?" are in need in schools (*BGIR, xvii*), the Committee models the underlying point of asking "which boys?" by covering Indigenous issues, socioeconomic issues, and family structures. Other differences were elided, though; sexuality is chief among them.

Third, the Committee critiques harmful media images of masculinity in their Chapter 3, though there are some conservative undertones present. They, like many (pro)feminist scholars with whom they spoke, decry the limited versions of masculinity the media offers. The Committee, though, avoided concrete recommendations to fix or counter these media images, despite (pro)feminist alternatives offered to them, such as explicit instruction in gender construction and critical media literacy.

The Committee also rejected several standard conservative arguments about boys as the "new disadvantaged." They rejected single-sex schools and classes as a panacea for boys' ills, reminding readers that *how* those schools and classes are conducted matters more (*BGIR*, 86). They also rejected any notion that single parents and women are to blame for boys' declines (54, 167). Indeed, even though the report stands as a largely conservative document, it can be seen as *moderately* conservative, for it rejects the most explicitly antifeminist arguments—from what Sawford called the "loopy feminist conspirators."[66]

Finally, another laudable feature of *BGIR* is the implicit call for whole-school and whole-child solutions to boys' problems. Whatever the political positioning of their "solutions," the attack is multifaceted, from phonics curriculum and pedagogy to public service campaigns, from teacher training on gender issues to vision and hearing screenings. Thus, one cannot criticize the report for being "single minded" or solely "tips for teachers."[67] Though their specific arguments are for conservative methods and structures, *BGIR* at least puts forward that the problem(s) of boys cannot be solved in a straightforward, simplistic way, but rather must be engaged on many fronts.

Elements of Bad Sense

The main worry about a national policy on boys' education, of course, is that it would hurt the gains of girls or take attention away from girls' remaining needs. At a micro level, little can be known about how each teacher has shifted his or her attention, but at a national level, much of the momentum for girls' education has already been lost to boys' education. Few major Commonwealth projects are being planned for girls' education, the *GEF* has been incrementally weakened over time (see chapters 2 and 4), and the bulk of the money for gender programs has gone to boys (chapter 4). As of this writing, it is too soon to gauge the impact of these shifts on girls, but previous research suggests that the likely outcome will not be positive.[68]

Another critique of *BGIR* is that it ignores social dynamics that underpin boys' difficulties, particularly masculinity, heterosexism, and racism. Though the report decries drunken driving, suicide, and so on, there are no recommendations that provide schools with social services or suggest awareness campaigns about the gendered nature of these social ills. Also, of course, the asocial nature of the report is evident in the wholesale overlooking of sexism and violence. The entire platform of most (pro)feminist and social justice groups, in fact, has been jettisoned.

To illustrate, a major issue facing Indigenous students arises from living in a racist society that treats them as inferior to Anglo-Australian students in curriculum, discipline, and employment. This partly explains why Australian Aboriginal people have some of the worst academic outcomes among all indigenous groups on the planet. *BGIR*, though, makes no recommendations concerning Indigenous students. As often seems to be the case, rather, Indigenous boys are used as the *justification* for working with all boys, but little in the policy attempts to *fix* the Indigenous simply issues. *BGIR* says Indigenous students' problems are too complicated for the Committee to solve (35).

Other programs designed to counteract the harmful effects of certain hegemonic forms of masculinity have also been completely overlooked in, or purposefully left out of, *BGIR*. In fact, the Committee, in criticizing the *GEF*, made it clear that they were hostile to the notion that boys themselves might need to change negative behaviors:

> [Our suggested] approach [to a recast *GEF*] casts the educational objectives positively for boys and girls as opposed to the negative approach for boys implied in most of the current policy material—for example, about boys not being violent, not monopolising space and equipment and not harassing girls and other boys. (68)

Though the Committee received evidence from numerous (pro)feminist academics and educators suggesting that these various problems stem from versions of masculinity that reject and attack all things feminine, the Committee completely ignored this evidence. In fact, the words "sexism," "sexist," and "bully" (and their variants) never appear in *BGIR*. The word "homophobia" occurs only once (60), but only in a long quotation, and it isn't the focus of the passage. Even the word "violence" most often appears only when used to show boys as *victims* of violence (101, 161) or victims of the indifference of society to the violence with which boys live (36, 60).

The Committee has in essence elided any responsibility that boys themselves have for the violence they commit and the oppression of other groups because of it. They suggest no programs for confronting these issues, and they make no recommendations to fix them.

Similarly, *BGIR* lays out an authoritative vision for what boys' education is and should be, yet, at the moments of recommending policy actions, the Committee suggests only surface changes—technicist interventions into pedagogy, curriculum, and assessment—rather than targeting core reasons for males' educational behaviors and attitudes. Thus the Committee, for example, recommends scholarships to attract more males into teaching, but this does little to address other reasons males avoid teaching, like its feminine valence.[69] Also, the Committee recommends phonics instruction as the preeminent method of teaching reading, though much research—and even testimony before them—suggests that boys' poorer results have much to do with the feminine valence of reading and the rejection of curriculum that threatens their construction of masculinity.[70] Without tackling these dynamics of masculinity construction, the symptoms of comparatively poorer reading and the avoidance of teaching cannot be ameliorated.

One may also criticize *BGIR* and the Committee for being ideologically opportunistic. Further proof of the report's politically conservative nature is that it attacks perennial conservative targets and advocates teaching practices popular among conservatives. Throughout *BGIR* the Committee criticizes teachers unions, as mentioned above, as well as education departments, teacher education, and left-leaning academics—key targets of conservatives worldwide. According to the Committee, these groups have "understated" boys' needs and teachers are not "adequately supported" by them (4, 61–62). The report also advocates phonics, a longtime conservative attack on progressive education more so than an attack against the actual method of whole language, though the Committee does avoid being overly stereotypical of whole language (*BGIR*, 110).[71] Qualitative research, often taken by conservative political advocates as "soft" or leftist research, was also a target. Mr. Sawford frequently complained of the lack of quantitative data on issues, particularly the "quantitative basis" for the *GEF* (e.g., *BGIR*, 65). This hostile attitude toward qualitative research mirrors policy efforts by conservatives elsewhere to limit research methods; the requirements for "scientific" evidence of "what works" in the U.S. *No Child Left Behind* and *Education Sciences Reform Act of 2002* policies are prime examples.[72] Finally, ideological opportunism is seen in the Committee framing any blame as a state problem—a new round in the perpetual federalism fights—for chapter 7 of *BGIR* complains *in advance* about funding threats and data inadequacies based on state-level faults.

As my final critique, the policy-report suffers from potential conflicts of interest related to the cycle of legitimacy described above. Put simply, they relied heavily on testimony from people who stood to gain a lot of money from concerted government attention to boys, including consultants, authors, all-boys schools, and grant-seeking academics and schools. One must be concerned over a kind of circuit of fealty (and payola) amongst the participating members of this cycle. Those loyal researchers who supported the government position and continue to do so will be richly rewarded in the new implementation scheme. The BELS money, for example, the program guidelines say, could be spent on "consultants," which many of the government positions' supporters happen to be—the Rowes, Fletcher, Biddulph, Lillico, and many others. A richly complex political economy is involved, and thus far it seems that the spoils of the "war against boys" in Australia (about A$30 million [about US$26.175 million]; see chapter 4) will be going to the more conservative forces that have *thus far* "won" that war.[73]

Implications for the United States and Elsewhere

I suspect that politicians and policy processes are remarkably similar in most countries, being vastly more human, informal, chaotic, and geared to the interests of the already powerful than most people would be comfortable knowing. Recent critical studies of education reform in the United States—Smith's *Political Spectacle and the Fate of American Schools* and McNeil's *Contradictions of School Reform* are stellar examples—certainly support this conclusion. Looking closely at the behind-the-scenes operation of policy processes, then, is crucial to resisting regressive boys' education reform and supporting progressive reform.

In the U.S. context, there has been little large-scale policymaking on boys, the reasons for which are enumerated in chapter 6. In what policymaking that has been done at local levels in the United States, schools and districts have fallen into some of the same traps as Australian educators. Popular-rhetorical texts dominate the debates, and these have created an oft-repeated discourse that favors explaining the "disadvantages" faced by boys as the fault of feminists or teachers who ignore boys' supposedly different brain configuration.[74] (Pro)Feminist discourses are largely overlooked, though in the U.S. context this has much to do with having very few high-profile forums to broadcast such discourses.[75] Consultants and those with vested interests to continue viewing boys as "oppressed" have dominated the conversation, and most of the concern has been expressed for white, middle-class, suburban boys; when other boys are mentioned, it is usually only to justify looking at *all* boys rather than to make substantive policies to alleviate, say, poverty, racism, opportunity deficits, or rural isolation.

For progressives in other countries, there are certainly opportunities to learn from Australia's example. My Coda describes several strategic considerations, but regarding the policy creation process described in this chapter, the approach to policymaking in all contexts must involve crafting a coherent message that is both understandable and persuasive to policymakers. If Edelman and Smith are correct that politics in the United States are often about spectacle over substance, then those who would resist dominant conservative notions about equality must learn to compete on these terms.[76] This is not to say that U.S., Canadian, and other scholars should lower standards on their scholarship, but they must adapt from academia to the very different terrain of policymaking if they are going to challenge simplistic and stereotypical notions of gender and equity.

Conclusion

BGIR, like most policy documents, does much defining—defining of terms, causes, solutions, government roles, the ideals of acceptable society, and so on. In fact, this defining leads to the double entendre in this book's subtitle, *Getting Boys "Right."* The Committee defines, for all to see, what a "boy" is, what constitutes a "right" or "proper" boy, and how this is best accomplished; in this regard, my title is based on the title of the focal policy, *Boys: Getting It Right.* These declarations about "right" boys are all political and ideological decisions, all *particular people's* values authoritatively allocated. These allocations, further, are largely conservative and "Right"-wing politically; in this regard my title pays homage to Apple's *Educating the "Right" Way.*[77] Yet such decisions are not monolithic, not automatically accepted by the "recipients" of policy, and no simple ascription of "good" or "bad" in a totalizing sense can be placed upon *BGIR.* Still, the operationalizing of the definitions of boys in policy initiatives and programs, discussed in the next chapter, represent another step away from progressive gender policy in education constructed in Australia from the 1970s to the late 1980s. The deconstruction of progressive policies has become part of a new construction of boy-focused policies and programs. I now turn to this "new" construction project.

Chapter Four
Means to an End: The Resulting Initiatives

Mary Lee Smith notes that a "policy" includes three parts.[1] The first two were discussed in my introduction: policies are, as Ball says, *texts* and *discourses*.[2] The third part of a policy, *instruments*, describes those "means to an end" that policymakers employ in implementing a policy and "allocating its values."[3] These include mandates, rewards, punishments, demonstrations, and program development that the state uses as means to accomplish its policies and ideologies, sometimes called policy "levers." Where the previous chapter explored *BGIR* as text and discourse, this chapter analyzes the *instruments* of Australian boys' education policy: the initiatives resulting from *BGIR*. This is where the ideological elements of the Committee's work come to life. This is where "report" becomes "policy."

While many micro-initiatives—small, local, or school-level programs—have sprung up informally partly because of *BGIR*, this chapter focuses on four larger initiatives, the formal instruments the government has used to "spread the word," if you will, about boys' education and what they want done about it. These include: (a) two iterations of a conference funded by the Department of Education, Science and Training (DEST); (b) a "recasting" of the national gender equity policy for schools; (c) a push for more male teachers; and, of most impact, (d) a best practices grant system called the Boys' Education Lighthouse Schools (BELS) program, which later spawned a professional development initiative, the Success for Boys (SFB) program. In addition to describing these resulting initiatives, I explore their effects. While these effects are evolving and continuing, several important dynamics set in motion by boys' policy in Australia must be understood. The tactics and strategies engaged have crucial implications for the future of gender and education policy in both Australia and places that may follow their example.

Boys to Fine Men Conferences

Newcastle City Hall in March of 2003 was abuzz with anticipation for the start of the third biennial Boys to Fine Men (hereafter, B2FM) conference. Federal attention had recently been paid to the issues of boys, including *BGIR* and Commonwealth money for the conference itself, and the media too had been giving exposure to the issue. Now it was time to really discuss boys. So the lights had been turned low in the cavernous City Hall auditorium, spotlights were trained on the stage, and the bustling and slowly settling crowd had been treated to fancy pastries and tea. This was going to be a feast in every sense for the attendees—researchers and practitioners alike.

There were the few obligatory opening speeches. Conference organizers from the University of Newcastle-Australia's Family Action Centre, a creator and purveyor of many so-called boy-friendly materials for practitioners and likely the first university in the world to offer a graduate certificate specifically in the education of boys, also gave a few speeches. Then it was time for the boy bands.

The Starstruck Performers, four teens with the popular "bed head" hairstyle featuring gel-induced spikes hit the stage first. The multicolored lights danced around, and so did they. Thereafter Anthony Snape, a young man with similar spiky hair and a smooth voice, came on stage to sing the conference song, "Boys 2 Fine Men." Having a boy band and a conference theme song at a professional conference—strange though it might be to some—seems representative of the entire conference on many levels. It demonstrates the dramatic lengths gone to in making it a multimedia event, drawing on every sense to reinforce for attendees the fun, feelings, and legitimacy of the occasion. One could easily and frequently get caught up in the constantly reiterated common purpose—boys are in trouble and need our help—and in the bodily pleasures of food, music, and conversation that lent a sophisticated and alluring (though temporary) lifestyle to this common purpose. It was a three-day immersion in the "structures of feeling"—Raymond Williams' term for the ineffable, unarticulated affective structures that are developed "*in solution*" as social experience is lived.[4] Put simply, free backpacks and water bottles, prawn puffs, and lots of other people complaining about boys and tapping their feet to the conference song helped make boys' education seem important, socially cohesive, and legitimate.

A complaint of people ignoring boys' educational needs was pervasive at the 2003 iteration of the B2FM conference. The use of this refrain mirrors other social movements' use of neglect, hardship, standing up for rights against heavy odds, and even oppression as rallying cries, even when, as in the case of boys, the group facing the "hardship" is actually socially dominant. The post–boy band speaker, Richard Fletcher, one of the influential figures noted in the previous chapter and member of the Family Action Centre at the University of Newcastle-Australia, the conference sponsor and organizer, demonstrated clearly the discursive identity work being done to construct boys' education as a virtuous fight against a tide of obstacles. Judi Geggi, Director of the Family Action Centre and emcee, introduced Fletcher as a "founder" of boys' education in Australia, "a lonely role for a long time." Not rejecting this heroic presentation, Mr. Fletcher then began speaking, marveling at the vast "progress" made in boys' education since the 1990s. The conference's very existence, he noted, was proof that "we've" finally "recognized" boys' problems after years of struggling to get boys on the agenda.

Another proof of the advancement of boys' education, according to several speakers, was *BGIR*. It had been released only six months earlier, and the initiatives that were to spring from it were just under way. *BGIR*, most importantly, had echoed many of the conservative views expressed by Fletcher and others participating in the conference. Geggi's introduction of Fletcher also cited *BGIR*, saying that Fletcher's struggles had been "validated by the Federal Government's recent Inquiry into Boys' Education." Kevin Wheldall, founder of the remedial literacy program MULTILIT (Making Up Lost Time In Literacy), which took up about two full pages in *BGIR* as an exemplar program for teaching systematic phonics, claimed in his presentation at the conference that *BGIR* "validated" his position too.

This notion of "validation" through *BGIR* creates a thorny circular argument, what I referred to in the previous chapter as a "cycle of legitimacy." The problem, again, with using the findings of *BGIR* to legitimate the findings of research in it is that *BGIR* has appealed for legitimacy *itself* by using those research findings. To phrase it metaphorically, the egg has claimed that it came first based on the evidence that it came from the chicken. The findings of *BGIR*, though, come with the government's imprimatur, which gives its conclusions a modicum of acceptability to most Australians. The Education and Training Committee (EDT), however, isn't made up of education researchers—only politicians, even if with teaching experience—and so it cannot authoritatively "validate" research findings in a scientific sense; it can only do so in an ideological or political sense. Thus, it is important to view researchers' and consultants' claims in a political-discursive light, one that views such claims as negotiations and advertisements of the self—as bids for legitimacy.

One shouldn't conclude that conference attendees were dupes to tempura appetizers and appeals to the authority of *BGIR*, though. The participants were not suffering a false consciousness that led them to accept wrong thinking. As Apple convincingly argues, people come to take up conservative arguments "because [such arguments] *are* connected to aspects of the realities that people experience.... there has been a very creative articulation [by conservative groups] of themes that resonate deeply with the experiences, fears, hopes, and dreams of people as they go about their daily lives."[5] For many attendees—most of them program coordinators, teachers, administrators, social workers, and others who work daily with boys—their experiences were practical, based in their jobs and their intimate connections to trying to help boys they care about. (There were also others who made it plain that they cared little about boys' education issues at all [see also chapter 5]. At one afternoon tea, a participant confided to me that he was just there for the day off!)

For those whose connection to boys' education was personal and practical, the two B2FM conferences I attended—in Newcastle in 2003 and in Melbourne in 2005—offered many opportunities to explore their work as practitioners. The 2003 conference explored "School and Community Partnerships" while the 2005 conference focused on content "From Practise to Practice."[6] These themes, and the sessions exploring them, follow the general trend to focus on practitioners, convincing them to commit to boy-centered pedagogy. The 2003 conference, for example, saw sessions on lessons learned from alternative schools, what women can do for boys, reading programs, social and psychological programs at specific schools, ways to involve fathers, anti-bullying programs, peer tutoring, and even crime prevention workshops. The 2005 conference had sessions on "Hot tips and strategies" for boys' literacy, identifying gifted boys, using literacy surveys, instituting safe driving programs, involving males from the community, building relationships with boys, and some describing various BELS projects. There were presentations geared to the academic study of boys' issues, but the major focus, particularly in the breakout sessions, was on practitioner needs.

Viewing differences between the 2003 and 2005 B2FM conferences evinces changes in the political arena for boys' education in Australia. In 2005, the gourmet crudités and open bar still fueled the multimedia barraged participants, but gone were the pleas to finally take boys' issues seriously; the massive funds and attention to the issue had made such legitimacy bids unnecessary. In their place were frequent exhortations to celebrate the positive initiatives for boys and frequent touting of the higher numbers of participants at the conference (reported to be about 950, up from 250 at the first conference in 2000). Gone, too, was the presence—albeit small in 2003—of (pro)feminist scholars. In their places were even more practice-oriented authors and consultants with even more to sell.

Practitioners' concern, importantly, has financed those with economic interests in promoting the notion that boys have problems and that they—the consultants—have solutions. The more people on board, the more products they can sell—books, kits, posters, videos, programs, professional development sessions, and conference registrations. This may sound too cynical, for many have undeniably real, valid concerns about actual boys. It would be naïve, however, to overlook the economy that's been created around boys' education and this economy's role in furthering the issue. While some progressives profit materially and in their reputation from debates over boys, looking at the high-priced surroundings of the B2FM conference, clearly the conservative factions were enjoying most of the economic benefits.

The B2FM conferences are major sites where the boys' education economy shows its wide diversity. First, products were on sale in the exhibition hall. In fact, watching the change in the exhibition hall from 2003 to 2005 demonstrates the explosive growth of the boys' education economy. Both the space allotted and the number of vendors filling it had grown, from two or three vendors in the corner of a lobby in 2003 to ten or more lining the walls of a large ballroom in 2005. Second, consultants gave teaser demonstrations of their professional development workshops as conference sessions, hoping that participants would want to hire the consultants to come to their schools or would want to attend regional training sessions. For example, the Rock and Water Program, a quasi-martial arts program from the Netherlands that stresses that boys and girls are inherently different, offers books, videos, and various full-day and multi-day professional development workshops; its creator, Freerk Ykema, gave multiple sample demonstrations at each B2FM conference.

It was nearly impossible to escape the commercialization during the 2005 conference. One session, in fact, billed as a "celebration reception," was actually a sales pitch for the materials newly published by the University of Newcastle and, not coincidentally, available for sale in the hallway through which everyone had to pass to get meals. There was even a contest going on, in which a member of the organizing committee looked for people doing good deeds during the conference (to model how participants might look for and reward the good in boys). The prize was a kit of motivational supplies, including a poster, stickers, motivational cards, and other tokens to use as classroom rewards. The kit, they reminded us during each contest update, was also on sale at the University of Newcastle table. Also for the first time in 2005, advertising for products was printed right in the conference program.

The conference itself has also been a lucrative endeavor. In 2005, between registration fees—A$649 (US$566) per person, totaling nearly

A$600,000 (US$523,500), considering the advertised 950 participants—and more vendors—starting at A$1100 (US$960) per table—one can see that the conference was making a tidy sum. Even so, as one could likely guess from my inclusion of the conference as a government initiative, direct Commonwealth monies also supported the conferences. DEST freely advertised its A$40,000 (US$34,900) grant to the conference in 2003.[7] The Commonwealth also sponsored the conference in 2005, contributing A$50,000 (US$43,625), though this time their contribution sum was not widely advertised.[8] Sponsorship also came from state ministries and the mining conglomerate Rio Tinto in 2003 and 2005. Clearly, the conference has attracted a great deal of money, and this has shown up in the opulence of the occasion, if not in the quality of the content.

Of course, it is too simple—even insulting—to assert a solely economic basis to the concerns expressed over boys. Researchers need, however, to begin to think theoretically about the conservative ascendance of boys' education issues and the functions these issues play in the social order and in policy. To understand the growth in boys' education as partly influenced by economic considerations isn't to deny the lived realities of practitioners or to impugn the characters of those who make money from it. Understanding the complexities that underpin growth and change in the social world, rather, demand it.

Beyond economics, and beyond the honest altruistic desire to help others, educational endeavors also have legitimacy needs and needs for the production of technical and administrative knowledge.[9] Regarding legitimacy, first, there are tremendous cultural pressures—such as the (pro) feminist resistance—that have inflected the direction, speed, and shape of boys' education movements, and changes to the B2FM conferences' program have occurred largely as a result of such pressures.

Perhaps chief among the visible changes was the increased presence of Indigenous leaders and those concerned about Indigenous students. The first Indigenous Boys to Fine Men Forum (an optional event for an extra A$77 [US$67] per participant) preceded the 2005 conference. The program included a dance performance by a group of Indigenous children; a panel discussing how to "grow up our young fellas to have a strong identity"; a panel of boys discussing "what it's like for them growing up as young indigenous men in Australia today"; a keynote by Chris Sarra, the nationally heralded and sometimes maligned principal of Cherbourg State School, a high Indigenous population school in Queensland; and three facilitated discussions about "success stories" for Indigenous boys based around the same "big issues" as the larger conference: identity, literacy, and learning.[10] Many of these same themes and concerns for Indigenous boys made their way into the full conference's program. Whereas no keynote or session titles in 2003 referred specifically to "indigenous" or "Aboriginal" issues (though some discussed these topics), by 2005 three sessions had reference to indigenous issues in their titles, one of the five keynotes and one breakout session featured the aforementioned Chris Sarra, one of the conference organizers was Indigenous and discussed indigenous issues in several talks to the main conference, and other Indigenous people were visibly participating. Even one of the vendors specialized in books by and for Indigenous people. This greater visibility of and focus on Aboriginal people can be seen as the product of many cultural movements and pressures, including the struggle by Indigenous people to be recognized by educational organizations and conferences. Initiatives like the B2FM conference and others run by the University of Newcastle (like their father involvement programs) also respond to the expressed needs of Indigenous people to implement community-based programs. Concerns remain that such initiatives are geared to assimilating Indigenous boys or that indigenous boys are being exploited as shills to get attention paid to all boys, but positive shifts are being made and, most importantly, are being led by Indigenous people.[11]

Other progressive pressures are evident in the changing conference program. While both conferences were heavily focused on literacy, involving men (particularly fathers), relationship building, and motivation, notable 2005 additions include combating homophobia, tackling boys' risk taking, and expanding boys' emotional repertoires. For the first time in 2005 a handful of sessions began to reflect awareness of the need to alter masculinity to fit pedagogy, not just the reverse.

Still, many—perhaps most—of the sessions in both iterations of the B2FM conference demonstrated strong conservatism. For one example, 2005's third day's opening keynote by Dr. Martine Delfos, from the Netherlands, evinced many conservative ideologies. She showed clips from a Belgian television documentary in which children were supposed to share a Viewmaster picture-viewing toy; the boys fight while the girls cooperate. Later she showed a lesson with clay; the girls build things while the boys just leave a mess. The simplistic conclusion presented is that boys and girls are just different, and biologically so—boys, she says at one point, cope with danger by "fight or flight" while girls cope by "nice or victim"—but that educators and researchers have interpreted these differences incorrectly, preferring the girls' behavior because it is "nicer." Delfos reinscribes conservative arguments that "boys will be boys," that there is a natural and biological basis for how boys act (what Delfos, in the program booklet, describes as "evolutionary, deeply embedded preference behavior" [12]), and that current education neglects or misunderstands these "common sense" "facts."

Crucially, though, conservative discourses are sometimes packaged alongside seemingly progressive, student-centered practices. Such contradictory impulses can, on first inspection, seem tense and dissonant, and it can even make the conservative tendencies difficult to discern. Yet these warring tendencies represent both the constant need for conservative impulses to moderate themselves and the continued viability of progressive movements in opposition to boy panics. One session by Richard Fletcher and Deborah Hartman in 2003, for example, outlined ten considerations for a "boy friendly school," including that it:

1. Knows how it is doing with boys and can tell you...
2. Has policies and practices based on a positive approach to male identities...
3. Uses teaching, learning and assessment styles that uses boys' strengths...
4. Has behavior policies and procedures to assist in positive identity and relationships...
5. Has school structures that are likely to suit boys...
6. Ensures that boys have access to a range of male mentors and models...
7. Is democratic...

   ...

   ...

10. Staff really try with boys and don't settle for "getting by"[12]

Many of these goals seem benign or even laudable, but much rightist ideology lurks within these recommendations. Number 2, for instance, recommends a "positive" attitude toward male identity, and number 4 calls for "positive" policies and procedures for "identity and relationships." Viewed in tandem with *BGIR* and Fletcher's submission to the Inquiry, the specific valence of that word "positive" becomes clearer.[13] The word is often used in opposition to a notion that boys are viewed negatively or their needs are only valid in so far as girls need them to change. Fletcher, for instance, in his Inquiry submission on behalf of the Family Action Centre, recommends that the Committee:

> ...Set out the principles under which a *Gender and Schooling* policy would be developed.
> a) A positive approach to boys and to meeting boys' needs is essential.
> b) The aim of strategies addressed to boys should be to develop fine men, not primarily to remove injustices suffered by women although this is a valid social goal and does imply change for males. (2)

BGIR, indeed, parroted this recommendation, saying

> This [recommended] approach casts the educational objectives positively for boys and girls as opposed to the negative approach for boys implied in most of the current policy material—for example, about boys not being violent, not monopolising space and equipment and not harassing girls and other boys. (68)

The same terminology of "positive" framing of boys policy finds its way, in fact, into the first recommendation in *BGIR*, where it is recommended that the *Gender Equity Framework* (*GEF*) be recast to "address boys' and girls' social and educational needs in positive terms" (*BGIR*, 69). "Positive," then, indexes key conservative arguments about the cultural stigma boy advocates perceive has been placed on boys.

Many of the messages of boys' education represented in the list of boy-friendly school characteristics, though, suggest educators should shoulder blame, guilt, or shame for their roles in creating such situations for boys, arguments that ought (intuitively) to be poor motivators for approaching reform, but clearly no difficulties have surfaced in motivation for educators to take up boys' education.[14] The participation of nearly 1000 educators and other practitioners suggests this strongly. So what about this message appeals to practitioners? Why does the conservative message not turn them off?

Part of the answer comes in the message's practice-oriented nature. It appeals to those facing real problems with boys, and, true, it suggests to them that they may be in part responsible, though unknowingly. It also, conversely and more importantly, *suggests that there are things they can do to fix it*. Such arguments often suggest, moreover, that the fixes are relatively simple and technical. Tweak the discipline system, alter some pedagogical strategies, and let boys have a say, and—POOF!—much will be better. Such strategies, of course, don't approach many core causes of boys' difficulties, including poverty, racism, violence, and patriarchal socialization. Even so, the fusing of progressive impulses (such as democratic participation) with conservative discourses provides a more palatable entrée into boys' education for practitioners, thus softening the conservative discourses.

Leaders in the Australian "what about the boys?" movement have obviously recognized the power of using a practitioner orientation to bring people into the fold. This is apparent in the explicit shifting of focus toward what works and away from despairing over problems. In the 2003 conference, most speakers (of those I saw) began with statistics that asserted boys had problems. In 2005, most presenters eschewed this and simply assumed the problems. Instead, the push in 2005 was to "celebrate successes" and show off "what works." In addition to the emcee entreating all to revel in successes, a flier in the conference program queried "Isn't it time we celebrated our successes?" and advertised a new University of Newcastle book by Hartman, *Educating Boys—The Good News: What Works for Boys and Teachers*. This positive shift is partly, of course, due to more general acceptance of conservative boys' education discourses. The presence this time of BELS schools and results from the first round of the BELS program (discussed below) also drive the shift, because more experience and data from practitioners were available. Also, this general trend toward touting "what works" follows other conservative movements worldwide—like the What Works Clearinghouse in the United States—which are moving toward "evidence-based research," though the state carefully controls what counts as "evidence," based often on ideological preference rather than scientific consensus.[15] The appeal relies partly on the pleasure of positive feelings being attached to an issue, a "structure of feelings" that may be as effective as or more effective than moral panic in rallying practitioners.[16] It also relies partly on the notion that things *do* work, and thus educators, social workers, and other practitioners can actually do something about their daily difficulties. Pragmatically and strategically for boys' education advocates, though, legitimacy demands that once you have political power and your way in policy, you can no longer complain but instead must show that your fix is working, at least for those "wise enough" to accept the plan wholeheartedly.

While the conference's major thrust clearly leans toward conservative understandings of boys, their problems, the problems' origins, and the solutions to them, part of the conservative discourse's success comes from the absence from or abandonment of the conference's rhetorical and physical space by progressive voices. The major scholars and activists from the (pro)feminist response tradition were hard to find at the B2FM conferences.[17] While a few were notably present as panelists in 2003 (including Bob Lingard and some high ranking femocrats), such scholars and feminist bureaucrats were conspicuously absent in 2005. Though in its current form it is not a conference that would appeal to such scholars and activists, B2FM without such voices was significantly devoid of counter-discourses that would provide alternative, progressive visions for boys' education to the hundreds of practitioners attending. As it turned out, (pro)feminist scholars would also eventually be absent from other key spaces in the debate, such as the creation of the new national policy on gender in schools.

"Recasting" *the Gender Equity Framework* (GEF)

BGIR's first, and perhaps central, recommendation is to "recast" the national policy that governs educational gender equity, the *GEF*. As mentioned in chapter 2, such reauthorization has been a required and recurrent activity, but the renewal process this time was to be different from the very beginning; this time it would be substantially more revision-oriented and, as I will explain, done by a more controversial group, one less dedicated to (pro)feminist ideals than any previously.

In *BGIR*, the Committee expressed concern, again, that the *GEF*, "built as it is on the prior work for girls, does not separately research and identify boys' needs and it sets boys' needs solely in the context of what still needs to be achieved for girls" (64). The Committee (really, mainly Mr. Sawford) also expresses concern that "witnesses, when asked, could not provide evidence of *quantitative* research to support the introduction of the 1997 *Gender Equity Framework*" (65; emphasis mine). Additionally, *BGIR* cited concerns that "The factors limiting boys' educational achievement do not exactly parallel those that affect girls" (65), contrary to what they claimed the *GEF* implied.

Based on these complaints, it is no surprise that the text of *BGIR*'s recommendation targeted all of these perceived shortcomings (69). The Committee recommends *separate* (though "parallel") policies for boys and girls, something that (pro)feminists struggled to prevent for many years because of the potential distraction from girls' needs.[18] The Committee also recommended "positive" address of boys, which is, as mentioned earlier, a political rather than empirical argument. They also recommended expanding the indicators used to judge disadvantage so as to recognize boys' thus far neglected needs.

DEST began working toward the *GEF* recasting in January of 2003, just two months after *BGIR* was released and before the official government response—which endorsed the recommendation without reservation—was released in June 2003. Recasting the *GEF* is the province of the Ministerial Council for Education, Employment, Training, and Youth Affairs (MCEETYA), the group of all state and Commonwealth heads of education, but it fell to DEST to create a tender to find a research group to do the recasting and write a new draft that MCEETYA would eventually have to approve.

The request for quotes for the *GEF*—the advertisement asking for bids to do the recasting—asked those submitting bids to complete four major tasks, including, first, "mapping" relevant policy and "best practices" documents from states, all with an eye to inconsistencies between the states and between the states and the *Adelaide Declaration*, the standing national goals for all schooling.[19] This first task also required that the contracted group conduct a "focused literature review" of national and international research that has a bearing on "how a student's gender intersects with other factors like socioeconomic status, indigenousness and geographic location, to impact on the educational and social outcomes of students in schooling."[20] Finally, the first task involved synthesizing "relevant elements of submissions and transcripts collected during the...Inquiry into the education of boys."[21]

The data collected from the first task were to be used in addressing the second task: producing a "critical review" of the existing *GEF*. The intent was to determine how the framework could be recast to address the needs of all students, "best align with recent and on-going developments in social and educational understanding," refine the "indicators of improvement" to include "relevant" social measures, and provide "practical utility to users and maximum benefit" to its consumers while still maintaining "flexibility for schools and communities."[22] One can clearly see the major tenets of *BGIR*'s critique of the *GEF* embedded here in the tender to find a research team, particularly *BGIR*'s concerns for expanding indicators and providing for practitioners.

The third task is arguably the most important, for the consultants were being asked to actually create a new framework text based on the critique done in task two. This represents a break from the previous polyvocal gender equity texts produced in consultation with various groups, as this time it would be largely the work of the consultants (even though they would collect relevant opinions from the states and select academics). Discussions with MCEETYA—the fourth task—would occur only at the end of the writing process. In revising the *GEF* before that fourth stage, the request for quotes asks that the framework "acknowledge and address *in positive terms* the complex educational and social needs of boys and girls in diverse communities," establish evaluation criteria, and target "schools and systems, education practitioners, parents and school communities" (2–3; emphasis mine), the latter being a not-subtle critique that the existing *GEF* is primarily for academics. Here again one can see the critiques from *BGIR* within the task set for the recasting, and the functioning of *BGIR* as a policy rather than "just" a report is visible, too, for the format and direction of the actual "policy" for Australia's gender equity had been predetermined by the shortcomings decided on by the Committee.

The choice of groups to recast the *GEF* would turn out to be as crucial to the direction of the new document as were the recommendations of *BGIR*. After the first round of proposals, DEST determined that there were only two credible bids, but that both were too expensive, so they allowed both to rebid. One group included numerous academics from across Australia, all based in universities and all with major, peer-reviewed publications on gender and education, and most had conducted large-scale projects for DEST previously. I was among this group, asked by the principal organizer, Martin Mills, to complete the synthesis of literature task—on which I had already published—and the analysis of documents from the Inquiry, a task that I was already nearing completion of for my dissertation.[23] Other members included Wayne Martino, Debra Hayes, Bob Lingard, Joanne Ailwood, Jane Kenway, Nola Alloway, and Julie McLeod. The other finalist group, GaiSheridan International Pty Ltd (GSI), included two aforementioned popular-rhetorical authors and consultants with much program implementation experience, Deborah Hartman and Richard Fletcher, along with two former DEST bureaucrats, Gai Sheridan and Robert Horne. These consultants had done no previous DEST studies, but Fletcher, as I showed in chapter 3, did have the Committee's ear and many of his views made their way into *BGIR*.

Ultimately, GSI was chosen to do the recasting. While this was disappointing to the group I was involved in, a comparison of the bids submitted shows that GSI's proposal (obtained later) was better aligned with the request for quotes's stated concerns—and therefore the stated goals of *BGIR* that underpinned the request. While the Mills proposal took a measured stance that would maintain the "which boys, which girls?" focus of the current *GEF* and included a list of member's publications spanning 16 pages, the GSI proposal spoke in the dominant language of conservative educational reform that spawned the boys' education policymaking originally, language replete with jargon of the current conservative Government's plans for schooling writ large, and no long bibliography. Whereas the Mills proposal could have looked from an outsider's perspective to be aimed at maintaining the (pro)feminist core of the current *GEF*, the GSI proposal took a tack more critical of the *GEF*, one that was more congruent with the Committee's and DEST's contemporary dislike for the *GEF*.

A few examples demonstrate the alignment of the GSI approach with the dominant conservative critiques of the *GEF*. The GSI proposal lists five major criticisms-cum-recommendations. First, they wrote, *"the framework needs a goal*—the outcome measure by which ultimately systems and schools judge their efforts to achieve gender equity" (Annex B, 1; emphasis original). This mirrored Minister Nelson's push for outcomes-based measurement, particularly for providing comparability and accountability—a mainstay of most conservative reform efforts worldwide. Second, GSI's proposal asserted that "a framework needs to be apt for its public," but the current framework, they say, "pitches mainly to professionals [rather than 'schools and parents'] and its style is discursive" (1). This melded, again, with the common complaint that the *GEF* is an academic document and, by implication, holds no practical use. The third criticism holds that the framework "needs to be underpinned by *a clear statement of principles and values*, written in plain English" (1). This aligned seamlessly with Minister Nelson's efforts to establish "Values for Australian Schooling," and also mirrored the Minister's frequent public complaint that education is filled with too much "jargon." GSI's call for "pointers for action" (Annex B, 1–2), fourthly, mirrors the practitioner orientation of the Government's strategy and urges "agency" for local districts, also a popular current push, seen in site-based management reforms that devolve traditional government responsibilities to communities. Fifth and finally, GSI suggests that the current framework has a problem with "authority and presentation." As they explain,

> The current Framework Document [sic] is very plain, and introduced simply by a covering letter from the then Chairman of the Gender Equity Taskforce. It does not carry the endorsement of the Ministerial Council [MCEETYA]. Such low key encourages the inference that the Framework is the preserve of gender education specialists. We suggest that, if the new Framework is to reach its desired audience, it will need to carry clear Council endorsement, and to be attractively produced.... A more modern presentation and adaptation in a variety of forms to address different needs of professionals, schools, communities and children may also be appropriate. (Annex B, 2)

Numerous appeals reside here, but one is to MCEETYA's vanity, both praising the weight of its endorsement and suggesting that high production value will ensure the lasting usefulness of the document. (An oft-mentioned concern throughout this research was that various documents not "sit on the shelf to collect dust.") Also, GSI positions academia's concerns as secondary (both to bolster their legitimacy and to fan the flames of popular antitheory sentiments) and reflects important conservative realizations that form matters, both in the aesthetics of the document itself and in the "teacher-izing" (or, as they say, "adaptation") of complex arguments.

Put simply, GSI's proposal likely excelled the Mills proposal in the eyes of DEST because it demonstrated a use of the dominant conservative language and because it understood better the push for practicality and accessibility. In other words, GSI painted their gender equity landscape in the style of the day. This isn't to suggest that the Mills proposal was incorrect in its approach or weak in its arguments. It was, in fact, a solid proposal based on much research evidence from some of the most respected scholars, nationally and internationally, in the subject area. Nevertheless, the recasting's approach was clearly fixed from the start by the request for quotes and it favored the conservative approach of GSI.

As to the economy of the recasting, DEST disclosed that it paid just over A$97,000 (US$84,632) for the review, though Senate documents show the actual cost was A$110,852 (US$96,718).[24] Mills and colleagues' second round application came in at just under A$78,000 (US$68,055). Thus, though no official reasons were given for choosing GSI over Mills and colleagues, based on the price difference and informal conversations with sources close to the decision making process, it is unlikely that cost was determinative; I suggest that the aforementioned discursive distinctions made the difference.

The choice of GSI was controversial to various bureaucrats and academics on several counts. The primary concern, particularly in (pro)feminist and femocrat circles, was that GSI has a demonstrated advocacy interest (not to mention economic interest) primarily in *boys'* education, with little record of advocating for girls. A conflict of interest was also of concern, for two of the members of GSI were, as mentioned already, former members of DEST, the body awarding the contract. Concern among many suggested that the members of GSI lacked the qualifications, such as advanced degrees and research experience, to undertake such a massive and important task. I admit to having been somewhat uncomfortable with the elitism inherent in such talk, some of which may have been "sour grapes" from some who had been turned down for the project. Still, it is true that the balance of experience with research for government agencies at state, national, and international levels lay with the Mills group. The members of GSI, particularly Hartman and Fletcher, have largely published in the popular-rhetorical tradition, relying on anecdotes for data and typically eschewing thorough use of research literature.[25] This inexperience shows itself in their recasting literature review (task 1), where they include numerous works in the bibliography they've not cited, list journal names in lieu of engaging the content of specific articles, and suggest—with scant evidence—that theories of social construction of gender (which they incorrectly conflate with the learning theory "constructivism") should be "modified" because the theories are "widely debated."[26] In fact, social construction is the dominant conceptualization of gender, both in scholastic circles and in established Australian policy. Social construction is hardly "widely debated." Such faults are not inconsequential, for GSI exports these claims as "the theoretical underpinnings of this [new] Framework," underpinnings that mark a significant but unsupported break from established gender equity policy and educational thought.[27] The interviews of experts, the literature review, the "mapping" of jurisdiction policies and best practices, and the critical appraisal of the *GEF* were completed in February 2004 per the revised timeline mandated by DEST. By April 2004 a draft of GSI's revised framework was circulated. Since then, no final framework has been released. As of July 2008, even many DEST officials involved with boys' education had no idea where the new framework was in the process.[28] While being stalled does not definitively indicate problems with the document per se, one could reasonably infer that MCEETYA did not find the results acceptable.

While no official statement was made about why no final policy has appeared, GSI's draft circulated in April 2004 provides clues, both to the problems getting past the left-leaning MCEETYA (all members are, at this writing, Labor Party representatives) and to the form the document may ultimately take.[29] The GSI draft, in line with their proposal, presents two overarching goals for equitable schooling (though they speak of it as one goal throughout):

> Australian schools will:
> - afford every girl and boy equitable access to all aspects of schooling; and
> - enhance the educational and social outcomes of schooling for both girls and boys, and bring them closer together, so that on leaving school young women and men have access to comparable employment and contribute through their family and civic life to a just and equitable society. *(i)*

While at first glance these goals seem perfectly reasonable and benign, actually the focus on "access" and "outcomes" represents a backslide from the existing *GEF*, for the new draft policy elides attention to social interaction and school culture—what happens between "access" and "outcomes." This shifts the debate from how homophobia, sexism, and other social problems create gendered disadvantage, moving it back three decades, back to the 1970s' questions of whether students have equal access to programs and get equal test scores.

GSI, in detailing the steps toward their vision of equity, lists five "guiding principles for actions to achieve the goal[s]," each of which they elaborate through the rest of the document. They summarize these, thus:

> Working in collaboration with their communities, Australian schools will
> 1. Support and challenge girls and boys to develop a positive and expansive gender identity and to understand that there are many ways of being masculine or feminine;
> 2. Promote positive and respectful relations between the sexes, and protect students from slight or harm on account of sex or sexuality;
> 3. Identify the common and distinct educational needs of girls and boys, and of specific groups of girls and boys, and address those needs through the curriculum, and methods of assessment and teaching;
> 4. Empower and challenge students to develop their career aspirations and approaches to family and civic life, free from gender prejudice; and
> 5. Prepare and implement a gender equity strategy, informed by the *Gender Equity Framework*, which covers:
>   - exercise of leadership in pursuit of gender equity, assignment of clear roles and responsibilities for achieving it, and the formation of effective partnerships with parents and relevant agencies in the community
>   - enabling staff to undertake professional development in line with the strategy; and
>   - regular monitoring and evaluation of the strategy and associated programs in line with the performance indicators in this Framework, so as to provide a basis of evidence for review and improvement. *(i)*

Several points should be made about this list. First, the language used demonstrates a subtle but important shift. The first principle's use of "gender identity," for example, as GSI explains in the introduction (2–3), is meant to signal a change to their "modified" theory of gender as a social construction. They "retain the fundamental point" that one's "gender identity," rather than "gender," is socially constructed, but they make clear that using such terms is meant as a rebuke to the 1997 *GEF*'s use of "power, in the form of dominant masculinity, as the main dynamic driving the social construction of gender" (2–3). Such a view of dominant relations of power underpinning gender in schools, they claim—without citation or evidence—"does not accord with the evidence of what is happening in schools, and is not generally accepted by teachers and students" (3). They substitute, instead, the idea that schools have a role in "assisting all students to develop a strong sense of themselves as girls or boys, within the context that there are numerous acceptable ways of being masculine or feminine" (3). The authors have, in essence, attempted to strip gender of its power relations, as if all genders and displays of gender were equally valued, rewarded, and represented in society. They also strip away the realization that certain people profit socially, politically, economically, and educationally from unequal relations of dominance.[30] This watered-down approach to social construction is ahistorical and apolitical at best, a backlash at worst.

The use of "slight or harm" in the second principle performs a similar task, erasing concepts like "oppression," "heterosexism," "privilege," and "disadvantage" that have been hallmarks of previous *GEFs*. "Slight or harm" devalues the often serious suffering of many students, and it again elides explicit power dynamics that often underpin such suffering. "Gender prejudice," rather than, say, "sexism," does similar discursive work in principle four.

Another feature of the language is pertinent: the frequent nonspecification of gender or the treatment of girls and boys as equal categories. First, this conveys the sense that boys and girls are located on an equal social level or at equal disadvantage and, second, leaves nearly every point open to interpretation as to who is in a position of disadvantage. Take, as just one example, GSI's recommendation that schools "encourage the development of positive and respectful relations between the sexes at school and in the future" (7). Whereas the overwhelming evidence is that males perpetrate violence and harassment more than females, GSI's implication is that there is equal need and equal responsibility.[31] They imply the same when saying, "girls and boys each have some distinctive needs, as well as many which they share in common" (7). While this is true, it leaves open the impression that each need, no matter how slight, is morally equivalent to all other needs and just as socially beneficial when they are met. In the end, this is the position that they must take, strategically, in putting forth a policy that can be taken up by both boys' advocates and girls' advocates. Ambiguous and softened language is required to accomplish such potentially politically combustible work and to reconcile the inherent ideological tensions of presenting the genders as equally disadvantaged.

A second key point, apart from diluted language, is that the recast framework's principles ignore the social and economic realms. It lays the responsibility for change in gender dynamics squarely with teachers and schools. Whereas the goals, they say, can be accomplished by altering "curriculum and methods of teaching and assessment," (20) they overlook social processes outside schools' control, namely governmental policy, corporate hiring and retention, structural inequalities that exacerbate gendered differences in the division of labor, and social barriers including sexism, racism, religious intolerance, and homophobia. The draft of the new framework sets no evaluative mechanisms to assess those factors and fails to suggest interconnections of extra-educational service provision or potential policy overlaps in other areas of society and governance (such as, say, industrial relations).

Third, and related to delegating full responsibility to schools, the draft framework attempts to devolve policy responsibility to the local level. As suggested by *BGIR* and the request for quotes for the recasting, the new draft suggests that local communities be involved in the negotiation and production of their own gender equity policies. While all of GSI's principles discussed above *imply* this because the list is introduced with "Working in collaboration with their communities, Australian schools will:" *(i)*, the fifth principle of the new proposed framework actually mandates it. Part of this impulse is commendably democratic. Still, the authors don't adequately address the possibility that schools will develop sexist rather than antisexist strategies based on what the community wants.[32] The new *GEF* authors also fail to consider that many schools, because of their already intense workloads, will likely rely on boilerplate policies ready-made by advocates and consultants—what I referred to previously, after Lingard, as "de facto policy."[33] Furthermore, no accountability measures are suggested for those schools that don't succeed or don't create equity plans, making this policy framework a largely symbolic, unenforceable mandate, one that can be reinterpreted to fit anything a community may want, no matter how regressive.

Further, the proposed framework skews toward boys' concerns, just as some feared when DEST awarded GSI the project. For example, in laying out important educational and post-educational outcomes for males and females (4–5), the authors provide examples of male disadvantage as most salient for five of the eight outcomes, and one of the remaining three suggests males and females are equal. Though a simple, decontextualized count of words indicating sex or gender (that is, "girl," "female," "feminine," "women," and their variants, along with the corresponding male terms) shows only four more male than female terms (197 to 193) in the draft, there are other imbalances, like weighting of lists, as just mentioned, and differing use of detail and evidence. Consider, for the latter, the following two paragraphs:

> Closer scrutiny shows that over time Australian girls have participated more strongly and with greater success in the higher levels of mathematics and

in the physical sciences. But there is still more work to do to motivate more girls to raise their sights in those subjects, and to move more confidently into technology studies, including computing.

Boys' attainment, especially in literacy, features strongly in public debate. In tests of reading literacy among 15 year olds in OECD countries in 2000, Australia's boys scored significantly above the OECD mean for boys, just as our girls were above the mean for girls. But the gap between them, though not unusually large in OECD terms, signified that on average our boys were roughly one school year behind our girls in reading literacy. Lower literacy feeds through to lower participation and success in school subjects requiring strong literacy, and lower schooling outcomes....[34]

The nonparallel treatment demonstrates the skew of the draft, for whereas girls' deficiencies are generalized and redress is suggested in generic ways, the literacy indicators for boys are quantified with research and the likely dire implications of these literacy indicators are explicit. The boys have been treated with some nuance and evidence whereas the girls' indicators are treated as sloganistic afterthoughts.

I don't want to give the impression that the proposed framework is a wholly flawed document. Indeed, the document succeeds on many accounts, and these deserve mention. First, despite questionable modification, the authors retain the general understanding that gender is constructed, particularly through social interaction for children at school. This is key to conceptualizing the gender system as changeable. Second, they place heavy emphasis on teachers as professional agents uniquely capable of determining the needs of their particular students through classroom and school research, rather than suggesting that all boys learn in sex-specific, biologically determined ways that can be met with prescribed pedagogical tools.[35] Third, they provide a user-friendly format, a recommendation they made themselves in their bid proposal. This includes limiting the number of goals and principles; using a practitioner- and parent-friendly vocabulary; providing a glossary of specialized terms; and including sections on relevant issues, "key strategies for action," "useful questions" that might guide practitioners' classroom research efforts, and (mostly) measurable "indicators of achievement" for schools and governments to gauge progress. They also reject a deficit approach and concentrate on "positive" elements for boys, a discourse that I have thus far been critical of but which holds promise to ensure greater participation of boys in the process of gender change by avoiding a pedagogy of blaming and guilting boys.[36]

The proposed framework also speaks, it should be noted, polyvocally with the voices of progressive discourses—vestiges of progressive victories that hold hope for (pro)feminist advocates. The draft framework retains, for instance, awareness that all boys and all girls are not the same (akin to the "which boys, which girls?" refrain) and that some students suffer tremendously because of their gender and the interaction of their gender with race, socioeconomics, and "geolocation" (rurality, remoteness, or urbanity). It also recommends, even when *BGIR* tellingly does not, that homophobia still be regarded as a crucial problem deserving programs and support structures.[37]

Still, whatever its positives, on balance the shape and progress of the *GEF* "recasting" process has followed a course of conservative backtracking from the historical focus on girls' education in Australian policy. While the remaining progressive elements suggest that this framework doesn't fully represent an "endgame" in girls' policy, its definitive shifts in attention, funding, and language suggest conservative ascendance within gender and education policy.[38] It remains to be seen, though, the extent of the shifts, for, as noted already, this framework has yet to receive MCEETYA's approval. Time will tell what discourses will finally win hegemony.

Male Teacher Initiatives

A major and oft-repeated recommendation of the new draft *GEF*, and that of *BGIR*, suggests recruiting and retaining more male role models—male teachers, specifically—particularly for primary students. Underpinning this recommendation is the vaguely defended belief that male teachers and role models provide social and academic development benefits to boys, specifically (*BGIR, xxii*). Nowhere, though, do they cite research that proves such assertions—*how* do role models "matter" and *in what way* do children "benefit"?—instead only forcefully asserting a "commonsense" notion of males' roles in boys' lives. This lack of corroborating evidence weakened DEST's ability to act on *BGIR*'s recommendations, particularly in trying to get the established sex discrimination laws amended, which I explain shortly.

BGIR includes four recommendations (18 through 21) designed to address reasons the Committee found that males don't choose teaching or that might provide incentive to more males to participate in schools. Recommendation 18 suggests that the Commonwealth increase teacher pay; 19 suggests expanding teacher education admission requirements to include more than academic success; 20 (the central focus of this section) recommends a scholarship program to forgive student loans in the Higher Education Contribution Scheme (HECS), equally divided between male and female recipients; and 21 recommends that "education authorities" use professional development and Web sites to promote male role models.

These recommendations clearly target the Committee's conclusion that males, as teachers, are not in schools because of poor pay and long-term advancement opportunities (*BGIR*, 158) and poor cost–benefit balance for teacher education (159), and, as "role models" and volunteers, are absent because schools have not sufficiently encouraged them to participate (164–165).

The Government response to *BGIR* (outlined in the Introduction), though, demonstrated an oddly mixed reception to *BGIR*'s recommendations. I don't mean "oddly" in a political, ideological, or legal sense, but from the sense that the government response largely *rejects* the recommendations, but Minister Nelson and DEST—the authors of the Government response—later attempted in every way possible to institute them.

The Government response to recommendation 20, providing HECS-free scholarships for equal numbers of males and females, provides perhaps the most complicated case. In the Government response, the Minister and DEST reject the recommendation outright, though they claim to be "sympathetic to [the] line of reasoning" that "more male teachers are needed because of the importance of providing good role models to boys" (23). Rather, they find the proposed solution ill-suited to fixing the underlying problems that discourage males from teaching. They outline three main rationales:

- it [the HECS-free scholarship scheme] is likely to have little impact on the gender balance among teachers in schools, because such scholarships would inevitably be limited in number and many would probably go to students (whether male or female) already committed to teaching;
- the evidence suggests that HECS [repayment amount] is not a major determinant in student choices [of fields of study]; and
- such scholarships would set an undesirable precedent, as the same principle could be applied to many University courses which have unequal gender representation. (23)

Alongside the reception of the rest of *BGIR* in the Government response, this rejection seems uncharacteristically lengthy and strongly worded. The response goes on to say that the Minister will work in other ways to encourage more males into teaching, but clearly the Government response views the scholarship scheme as a nonstarter.

The strength of the objections to the scheme is also curious because they have contradictory place within the political timeline of the issue (see the timeline in the Appendix). On February 27, 2003, about four months after *BGIR* was tabled, Minister Nelson published a media release supporting the "need" for more male teachers and announcing that he had written all the deans of education at Australia's universities to ask what could be done. *The Age* newspaper reported on March 16 that Nelson got his answer from the Australian Council of Deans of Education: the scholarships are "simplistic and would fail to persuade men to become teachers."[39] The Council of Deans called instead for "improved career paths and radical changes to schools."[40] The Government response, as I said, largely echoed these concerns. Nelson, however, had already, on May 19, requested a review of the Sex Discrimination Act of 1984 (hereafter, SDA) to find ways to avoid it or change it to allow for "male-only scholarships."[41] Further, on May 3rd of the next year—2004—Nelson proposed a specific piece of legislation, entitled "Sex Discrimination Amendment (Teaching Profession) Bill 2004," that would both amend the SDA—an idea rejected by the opposition parties—and establish a A$1 million (US$872,000) fund for 500 male-only primary school teaching scholarships of A$2,000 (US$1745) each.[42] The small sum of each scholarship clearly signals the scheme's largely *symbolic, political* intent; it was more about changing the SDA than easing male students' burden with educational costs.

Much of the wrangling over the SDA was prompted by the Catholic Education Office in the Archdiocese of Sydney, which applied in 2002 to the Human Rights and Equal Opportunity Commission (HREOC) to have an exemption to the SDA so that they could provide 12 male-only teaching scholarships over five years. The HREOC, on March 3, 2003, rejected the proposal, citing *BGIR* numerous times *as support*. In general, the HREOC found that the lack of male teachers was about reasons broader than a scholarship scheme could remedy, including the status of teaching, child protection, and pay and conditions compared to other occupations.[43] The Commission, in fact, praised *BGIR*'s recommendation 20 because the scholarships were to be given equally to males and females. (Indeed, when the Catholic Education Office revised its application to give equal numbers of scholarships, the HREOC, in March 2004, approved it.) Minister Nelson, in a press release the day that HREOC turned down the initial proposal, expressed that he was "disappointed" by the Commission's decision. He used this disappointment to forward changing the SDA, despite the HREOC's many objections and even his own Government's response to *BGIR*.

Seen in the context of the *political* calendar, the contradictory and, some might say, bullheaded pursuit of a rejected scholarship policy makes more sense. Two thousand and four was an election year, and much of the wrangling over the SDA took place in the political—not legal—realm, for it was already known that Labor and the smaller parties were going to reject changes to the SDA. They had declared so in their dissenting report to the Coalition-dominated Senate Legal and Constitutional Legislation Committee, which had been tasked to review the proposed SDA amendment.[44] The debate, then, was clearly being used as a "wedge issue" for the upcoming election.

Evidence of the political impetus was plentiful. First, the rapidity with which the legislation was introduced was unusual. Previous amendments to the SDA have taken as much as four years to be enacted.[45] Also, the revival of the amendment followed soon after the initial campaign speech in mid-February by Leader of the Opposition Mark Latham, Prime Minister Howard's opponent in the election, in which, as described in chapter 2, Latham declared a "crisis in masculinity" driven partly by a lack of mentors for boys.[46] The speech got much favorable press and spurred an upward climb in the polls for Latham. Proposing this amendment to the SDA offered Howard a way to force Latham into the appearance of contradicting himself, for on its face the amendment offered one approach to attracting more male teachers. It necessitated changing the SDA, though, a move that, if Latham had agreed, could have alienated (pro)feminists, who would largely support Labor. This amendment was also clearly political in that it attracted such high-level public commentary. This commentary was tellingly directed at Latham, not at the core need for the amendment. As Prime Minister Howard himself said in Parliament:

> The reality is that the Leader of the Opposition, having run around the country and expressed his concern about male role models, being presented with an opportunity to do something practical, has run away from that.[47]

Prime Minister Howard also made similar charges in the media.[48] The language in this quotation—its focus on the person, not the policy, and the mirroring of his charge that Latham wanted to "run away" from Australia's participation in the second Iraq war—provides yet more proof of the political nature of the amendment and its purposeful timing.

As final evidence of the political rather than legislative intent, consider that after the election the pace of the amendment's progress slowed considerably. Despite a Coalition supermajority (that is majority in both the House and the Senate), the proposed amendment languished unenacted, out of public sight and out of ministerial media releases for (at this writing) nearly four years.

The Boys' Education Lighthouse Schools Program and Success for Boys

The BELS program, even more than academic research produced for DEST, has been a central engine for the manufacture of knowledge about boys' education in Australia, particularly practice-oriented knowledge.[49] The program is a takeoff on England's "beacon schools" program, in which schools were to be exemplars of best practice, showing—according to the metaphor—other schools the way toward good results.[50] ("Lighthouse," though, strikes me as a poorer metaphor since a lighthouse is supposed to warn ships *away!*)

The Australian BELS program was created in 2002, just after *BGIR*'s release, as a way to build a research base for promising ideas and exploratory projects. DEST structured the program in two stages, together costing just over A$8 million (US$6.98 million). In the first stage, conducted in 2003 over six months, 110 projects involving 230 schools received grants of between roughly A$5000 (US$4363)—depending on the amount requested—for individual schools and a maximum of A$20,000 (US$17,450) for four or more schools working together. Stage two would entail a smaller set of schools—initially intended to be 30 but eventually expanded to 51 schools—that would continue a successful program and would share their expertise with other schools. The grants for the second round were for A$100,000 (US$87,250) each. The Curriculum Corporation (the MCEETYA-owned educational materials producer) and Professor Nola Alloway of James Cook University were given the contract to administer the first stage, and the second stage was administered by Melbourne University Private under the direction of Professor Peter Cuttance. Schools in the first round were asked to conduct their interventions and to report their results by September of 2003, and by December a full report and an abridged "summary report" were released.[51] The second stage of BELS concluded in 2005 and was reported on in 2007.[52]

The first stage full report summarized the various projects conducted under the categories of "pedagogy, curriculum, and assessment;" "literacy and communication skills;" "student engagement and motivation;" "behavior management programmes;" and "positive role models for students."[53] In addition to concluding that schools work best in clusters, the stage one report's authors synthesized a list of ten "Guiding Principles for Boys' Education":

1. Collect evidence and undertake ongoing inquiry on the [boys] issue, recognising that schools can do something about it.

2. Adopt a flexible, whole school approach with a person and team responsible.
3. Ensure good teaching for boys, and all students in all classes.
4. Be clear about the kinds of support particular boys require.
5. Cater for different learning styles preferred by boys.
6. Recognise that gender matters and stereotypes should be challenged.
7. Develop positive relationships, as they are critical to success.
8. Provide opportunities for boys to benefit from positive male role models from within and beyond the school.
9. Focus on literacy in particular.
10. Use information and communication technologies (ICTs) as a valuable tool.[54]

Clearly, this list didn't deviate significantly from the conclusions of *BGIR*. One might conclude that the congruence between the two results from the Inquiry Committee being correct in their assertions, proven by the on-the-ground research of the 230 BELS schools. One might alternatively conclude that the congruence represents the repackaging of a priori conclusions. Consider that DEST-appointed persons selected the schools, filtered and abstracted the schools' reports, and authored the final BELS evaluation. The likelihood of a report standing in sharp contrast to already articulated DEST-accepted positions was small indeed. This isn't, moreover, the only evidence of the political dimensions of the BELS program.

Political economy: The spoils of the "war against boys"

A historical truism is that the winner of a war reaps the spoils. This appears to be true of the "war against boys," as well.[55] The "turn" to boys described in chapter 1 has not solely been discursive. Material resources, too, have become the booty of those who are winning the boys' education debates, as I detail throughout this chapter. Perhaps no other resulting initiative of *BGIR* has shown this more clearly than the BELS program.

What has the vast sum given under BELS—over A$8 million (US$6.98 million)—been spent on? Pinpointing the specific recipients of the grant schools' money is nearly impossible, but one can reconstruct from the reports the sorts of things schools did with funds. The guidelines for the BELS program suggest "funding can be used for such elements as: consultants, teacher release, direct delivery of specialist teaching and learning programs and professional support—including teacher professional development."[56] This continues the circuit of fealty and payola alluded to in the previous chapter, for those loyal researchers and consultants who supported the government position and continue to were, and will continue to be, richly rewarded in the new implementation schemes that allow schools direct funding. Progressive consultants also capitalized on BELS funding, of course, with (pro)feminist scholars engaged by schools to give cautions and action research advice (see chapter 5).

More than just money followed the BELS program, however. The spoils of educational policy wars frequently involve *political* capital too. With the BELS program, the engineering of politics could hardly be clearer.

Let me use as an illustration the initial allotment of 38 BELS phase two schools (another 13 were added later).[57] Table 4.1 lists the schools selected for funding and their locations. When one adds information about the schools' political electorates, a more interesting political dynamic emerges: *the selection of Lighthouse schools looks to be decidedly political rather than by chance or solely the merit of the schools funded.* My analysis shows that electorates represented by a member of the then-ruling Liberal-National Coalition had 25 of the 38 phase two schools. This represents 65.8 percent of phase two schools, though Coalition members made up only 54.7 percent of the members of the House of Representatives (82 of 150 seats). The Australian Labor Party (ALP), the left-wing opposition party, was only given 11 of the grants (28.9 percent) in its electorates, disproportionate to their numbers in the House (43.3 percent). A chi square test of the probability of this proportion of schools being allotted by chance shows that the differences are statistically significant ($p=0.0187$) and not likely random.

Admittedly, there is a statistical chance that the allotment of schools was an anomaly, that random (or meritocratic) selection distributed the schools in a skewed way. Indeed, with boys' education being dominated by politically conservative movements, one might expect that those electorates that would be most concerned about boys' education would also elect conservative candidates. Other indications, though, support a political explanation for school selection. First, as table 4.2 shows, the number of schools selected in each state and territory closely correlates with the state's relative population. The fierce rivalries between states certainly play a part in ensuring that each is distributed BELS schools according to their size and contribution to the tax base. Another politically sensitive consideration also demonstrates the deliberateness of the

Table 4.1 Phase Two Boys' Education Lighthouse Schools, by State, Educational Sector, House of Representatives Electorate, Member of Parliament's (MP) Party, and Seat Status

School name	Government/ non-government	Electorate	MP at selection[a]	MP party	2004 seat status[b]	2004 election result[c]
Victoria (n=8)						
Bayswater Primary School	GOVT	Aston	Pearce[e]	Liberal	Fairly Safe	Held
Brighton Grammar School	NON	Goldstein	Kemp[d,e]	Liberal	Fairly Safe	Held[f]
Chaffey Secondary College	GOVT	Mallee	Forrest	National	Safe	Held
Flora Hill Secondary College	GOVT	Bendigo	Gibbons	Labor	Marginal	Held
Hampton Primary School	GOVT	Goldstein	Kemp[d,e]	Liberal	Fairly Safe	Held[f]
Kew Primary School	GOVT	Kooyong	Georgiou	Liberal	Safe	Held
Pakenham Consolidated School	GOVT	McMillan	Zahra	Labor	Marginal	Lost to Liberals
St Joseph's College	NON	La Trobe	Charles	Liberal	Marginal	Held
New South Wales (n=11)						
Airds High School	GOVT	Macarthur	Farmer	Liberal	Fairly Safe	Held
Balgowlah Heights Public School	GOVT	Warringah	Abbott[d]	Liberal	Marginal	Held
Collector Public School	GOVT	Hume	Schultz	Liberal	Fairly Safe	Held
Kingscliff Public School	GOVT	Richmond	Anthony[d]	National	Marginal	Lost to Labor
Lismore Public School	GOVT	Page	Causley	National	Marginal	Held
Macleay Vocational College	NON	Lyne	Vaile[d]	National	Safe	Held
Oak Flats High School	GOVT	Throsby	George	Labor	Safe	Held
Quakers Hill Public School	GOVT	Greenway	Mossfield	Labor	Marginal	Lost to Liberals

Continued

Table 4.1   Continued

School name	Government/ non-government	Electorate	MP at selection[a]	MP party	2004 seat status[b]	2004 election result[c]
St Andrew's College	NON	Greenway	Mossfield	Labor	Marginal	Lost to Liberals
St Joseph's Primary School	NON	Hunter	Fitzgibbon	Labor	Safe	Held
West Wallsend High School	GOVT	Charlton	Hoare	Labor	Fairly Safe	Held
Queensland (n=6)						
Blackheath & Thornburgh College	NON	Kennedy	Katter[e]	Independent	Safe	Held
Dimbulah State School	GOVT	Kennedy	Katter[e]	Independent	Safe	Held
Mirani State High School	GOVT	Dawson	Kelly	National	Fairly Safe	Held
Thornlands State School	GOVT	Bowman	Sciacca	Labor	Marginal	Lost to Liberals
Urangan State High	GOVT	Wide Bay	Truss[d]	National	Fairly Safe	Held
Walkervale State School and Kepnock State High	GOVT	Hinkler	Neville	National	Marginal	Held
Western Australia (n=4)						
Melville Senior High School	GOVT	Tangney	Williams	Liberal	Fairly Safe	Held[f]
Merredin Senior High School	GOVT	Kalgoorlie	Haase	Liberal	Marginal	Held
Moerlina School	NON	Curtin	Bishop[d]	Liberal	Safe	Held
Tom Price Senior High	GOVT	Kalgoorlie	Haase	Liberal	Marginal	Held
South Australia (n=5)						
Mount Barker High School	GOVT	Mayo	Downer[d]	Liberal	Safe	Held
Salisbury High School	GOVT	Wakefield	Andrew	Liberal	Marginal	Held
St Augustine's Parish School	NON	Wakefield	Andrew	Liberal	Marginal	Held

Continued

Table 4.1 Continued

School name	Government/ non-government	Electorate	MP at selection[a]	MP party	2004 seat status[b]	2004 election result[c]
Open Access College R-10	GOVT	Sturt	Pyne[d]	Liberal	Fairly Safe	Held
Youth Education Centre	GOVT	Adelaide	Worth[d]	Liberal	Marginal	Lost to Labor
Tasmania (n=2)						
Claremont College	GOVT	Denison	Kerr	Labor	Safe	Held
St Thomas More's School	NON	Bass	O'Byrne	Labor	Marginal	Lost to Liberals
Northern Territory (n=1)						
Kormilda College	NON	Solomon	Tollner	Country Lib	Marginal	Held
Australian Capital Territory (n=1)						
Palmerston District Primary School	GOVT	Fraser	McMullan	Labor	Safe	Held

Notes:
[a] Phase two schools were announced in April 2004. [b] Classifications accessed February 25, 2005, from http://vtr.aec.gov.au/DivisionalClassifications-12246-NAT.htm. [c] The election was held October 9. Final results retrieved February 25, 2005, from http://vtr.aec.gov.au/SeatsWon-12246-NAT.htm [d] Minister or Parliamentary Secretary at selection. Shadow ministry not included. [e] Member of the Committee during boys inquiry (39th or 40th Parliament). Mr. Kemp commissioned the *Inquiry*, so I have included him. [f] Incumbent retired or didn't run.

Table 4.2 Number of BELS Phase Two Schools, by State, Percent of Total BELS Schools, and 2001 State Percent of National Population

State	n	Percent of BELS Schools (n=38)	Percent of Australian population
New South Wales	11	28.9	33.7
Victoria	8	21.1	24.9
Queensland	6	18.6	15.8
Western Australia	4	10.5	9.8
South Australia	5	13.2	7.8
Tasmania	2	5.3	2.5
Northern Territory	1	2.6	1
Australian Capital Territory	1	2.6	1.6

allotment of BELS grants: public versus private schooling. There is a high degree of correlation between the allotment of grants to government schools (28, or 73.7 percent) and nongovernment schools (10, or 26.3 percent) according to their share of the student population (in 1999, 72.7 and 27.3 percent, respectively).[58] Had more nongovernment schools than their "share" been awarded grants or vice versa, a firestorm of protest would surely have ensued. If the allotment of schools could be influenced and constructed by state and sector, it doesn't stretch credibility to think they might also be chosen partly by political party.

What are the material implications of a politically oriented selection of these schools? Why do it? Certainly, for the schools, such an infusion of funds would galvanize support for the project. For politicians of the ruling party, the awarding of funds within their electorates provided an opportunity to tout the existing government's concern for education and children, as well as a concrete demonstration of their largesse when social problems were identified. It was a vote getter.

Perhaps all this sounds too cynical. Sure, one might think, there is a political benefit of giving a local school a big grant, but that's hardly proof of politically engineering the selection process. Consider, then, that the selections were made only six months before a federal election. That the selection of schools would be politically apportioned in an election year wouldn't be unthinkable. The evidence, in fact, suggests just that. Consider the electorates, again, that were apportioned BELS schools (table 4.1). In all federal elections, the Australian Electoral Commission classifies each electorate by its relative "safety," meaning the chance that the incumbent party will hold the seat in the next election. There are three categories: "safe," for those electorates where the incumbent polls 60 percent or higher; "fairly safe," for those polling between 56 and 60 percent; and "marginal," for those electorates where the incumbent polls at less than 56 percent. The latter, marginal seats (often called "battleground" electorates in the United States) could potentially go to either party. Because such large BELS grants would be a vote getter, selecting more of these marginal seats to receive BELS schools would be a help to the ruling Coalition. As table 4.1 indicates, 18 of the BELS phase two schools were granted in marginal districts, ten in fairly safe electorates, and ten in safe electorates. Using a chi square test to compare observed and expected values, there is only a low probability ($p=0.1009$) that those numbers resulted from chance. Only 34.7 percent of all 150 seats in the House of Representatives are considered marginal seats, whereas 47.4 percent of BELS schools were apportioned to marginal electorates. It would certainly benefit the Coalition to have more grants go to those electorates where their members are struggling to keep or gain seats. As it turned out in the election, the nonincumbent Coalition candidate won four of the marginal electorates given BELS schools. Nonincumbent Labor candidates won only two.

By pointing out that the selection of schools was political, I don't mean to suggest that schools that won the grants were otherwise undeserving. Certainly there are significant and effective programs amongst these schools. I do suggest, however, that merit *alone* may not have been the sole deciding factor. Location (by state), sector (government or nongovernment), and—I argue—politics had a great deal to do with it too.

Positive and Tentative Results

Despite its *partial* genesis in political gamesmanship, many positives have come from BELS. Indeed, much research has been conducted under the program, determining numerous effective and progressive practices and warning against some regressive practices. Within the ten guiding principles for boys' education from the phase one report, there are progressive sentiments, particularly those pointing to the systemic and social nature of boys' difficulties and the need to disaggregate boys for particular targeting.[59] Numerous programs were developed that targeted boys who were in serious need of help—"students from challenging socioeconomic circumstance, linguistically and culturally diverse backgrounds, Indigenous students, students with disabilities and students deemed 'at risk' "[60]—and several programs showed extremely positive results. The Macquarie University Special Education Centre, for example, used the Making Up Lost Time in Literacy (MULTILIT) program to increase reading accuracy and comprehension in boys by more than a developmental year on average.[61] Southwell Primary School, too, improved attendance, decreased disciplinary referrals, and improved literacy measures by instituting a program to help Indigenous boys who were deemed "at risk."[62]

Even with some successes, the true usefulness of the BELS program remains to be determined. Its short-term nature (phase one was completed in just six months) raises the possibility of results being due to the Hawthorn effect rather than stable and predictable gains. Thus, for example, taking boys camping may improve attendance and behavior in the short term, but does this activity have impact longitudinally? Additionally, much of the information collected from the schools in the first stage was highly anecdotal and rarely quantified. The second stage appeared to fix some of these deficiencies, though. The second stage program administrators frequently urged schools to "gather evidence to show whether your innovation has improved learning outcomes for your students," and the project was extended, this time lasting a year and nine months.[63] The result was a report with stronger evidence of real successes.[64]

Success for Boys

Success for Boys (SFB) might be viewed as a third stage of BELS, for SFB continues most key features of BELS, though with some important differences. First, SFB is foremost a teacher professional development program rather than student development program, with plans to grant up to 1600 schools about A$10,000 (US$8725) each to "purchase and implement professional learning in boys' education."[65] This "professional learning" refers specifically and solely to a SFB curriculum package that provides professional development in five areas: a "core module" that addresses boys' issues "holistically," and one module each on boys and literacy, male role models, information and communication technology (ICT) as aids for engaging boys, and Indigenous boys' transitions from primary to secondary school.[66] The conduct of this program suggests that the Government may be taking tighter control of the framing of boys' education initiatives by codifying it in professional development units rather than having schools develop their own research. The Government also escalated the funding. Whereas the two stages of BELS came in at around A$8 million (US$6.98 million), SFB is slated to cost A$19.4 million (about US$16.93 million) through 2008, inclusive of the materials production, the pilot trials, and the A$16 million (US$13.96 million) in direct grants to schools.

There are progressive possibilities of SFB just as there were in the BELS program. The presence of progressive scholars on the development team, including Nola Alloway, a well respected feminist, firstly, represents a positive change from previous initiatives. Secondly, focusing on Indigenous students' specific needs is a victory, clear evidence of the successful struggle to focus attention on the most disadvantaged boys rather than all boys. The ultimate results of the SFB program will become clearer as it progresses over its expected two and a half years, through 2008.

Implications of the Resulting Initiatives of *BGIR*

Almost no policy change can be isolated, without any impact on other areas. Governments, usually aware of this, tend to carefully consider using new policies and programs, often either trying to make them fit prevailing ideological, political, or economic conditions or, perhaps more often, trying to use policies and programs to shift prevailing conditions to better fit their ideal views of society, government, or life. Thinking through the kinds of changes that *BGIR* and its resulting initiatives might make is my task in this final section. Just what are the implications for the ecology of Australian schooling?

The Economy of Boys' Education

Perhaps the greatest impact the Commonwealth government has had on boys' education has been the infusion of massive funds, thus *creating* an economy around boys' education. While the B2FM conferences demonstrate that an economy has grown from practitioner needs, the Commonwealth has helped *create* this economy to a degree greater than most other actors in the debate. Commonwealth funds have made research possible, just as they've made it possible, through grants, for schools to purchase the fruits of such research: materials, books, and expert consultants. As table 4.3 shows, boy-focused initiatives have attracted much Commonwealth money, at nearly A$30 million (just over US$26 million), far outstripping Commonwealth funds for girl-specific programs during the same period. What is also apparent from this roughly chronological list is that funding for boys' education has escalated dramatically, from the moderate first stage of BELS to the lavish spending on SFB. The list also evinces a shift from funding going to progressive scholars to funding going to conservative consultants, with a correlated shift in the kinds of research funded.

The creation of an economy around boys' education has many potential implications. Perhaps of most concern is the amount of control the Commonwealth asserts over schools, programs, and researchers that accept funding. The state's invention of an economy both creates and sustains *particular versions* of "truth." Just as federal intervention in the United States has greatly bolstered the fortunes of phonics within the reading wars, federal intervention in Australia has privileged the view that boys are in crisis and need attention. Those who accept that view receive money for BELS projects, get research contracts from DEST, are allowed to consult in schools, get books published to help teachers, and get promotion, tenure, and accolades based on these opportunities. With so many schools and researchers getting grants and producing research and procedural knowledge that supports the position they were initially paid to support, legitimacy circulates back to the government. In many ways, then, the government has literally purchased its own legitimacy. The same would likely have been true had the government supported the (pro)feminist position. Interests in capitalist societies follow money and vice versa.

Table 4.3 Declared Commonwealth Government Funding of Initiatives on Boys' Education, 2000–2005

Program	Funding allocated or spent (A$)[a]
Research: *Addressing the Educational Needs of Boys*[b]	244,800
Research: literacy development in boys[c]	212,200
Boys to Fine Men Conference 2003 sponsorship[d]	40,000
Newcastle Boys' Literacy Project[e]	44,000
Boys' Education Lighthouse Schools, stage one, direct grants to schools[e]	860,000
Boys' Education Lighthouse Schools, stage one, administration[f]	1,323,000
Boys' Education Lighthouse Schools, stage two, direct grants to schools[g]	5,100,000
Boys' Education Lighthouse Schools, stage two, administration[h]	699,500
Research: Mapping of gender-specific and -related curricula[h]	56,100
Gender Equity Framework recasting[h]	110,900
Male teacher scholarships[i]	1,000,000
Success for Boys (including trial of materials)[j]	19,400,000
Research: pilot study on father involvement[k]	50,000
Research: *Motivation and Engagement of Boys*[l]	338,200
Male Teacher Support (MATES) project[l]	133,600
Boys to Fine Men Conference 2005 sponsorship[m]	50,000
TOTAL	29,662,300

Notes:
[a] Values are rounded to the nearest hundred dollars.
[b] Commonwealth Department of Education Training and Youth Affairs, *Annual Report 2000–01*.
[c] Commonwealth Department of Education Science and Training, *Annual Report 2001–02*.
[d] Ministerial media release, October 21, 2002.
[e] Commonwealth Department of Education Science and Training, *Meeting the Challenge*.
[f] Commonwealth Department of Education Science and Training, *Annual Report 2002–03*.
[g] Ministerial media release, June 10, 2004.
[h] http://www.dest.gov.au/NR/rdonlyres/16DB0588-5B78-4728-968D-C65BD29CECA2/4057/8apr05_compiled_murray_order.pdf (accessed June 7, 2008).
[i] Ministerial media release, May 3, 2004.
[j] Ministerial media release, June 22, 2004.
[k] Ministerial media release, June 24, 2004.
[l] Commonwealth Department of Education Science and Training, *Annual Report 2004–05*.
[m] Personal correspondence with DEST.

Eroding Protections

In establishing boys' education as a legitimate concern, one worthy of funding, the Government had many barriers to cross, particularly the infrastructures and laws in place to protect and advance the rights and interests of women. The existing *GEF* militated against the sole focus on boys in pedagogy and curriculum just as the existing SDA militated against providing male teacher scholarships. For the former, *BGIR* suggested the *GEF* be changed and the Government responded by funding that initiative and by choosing a group of advocates intent on doing so. For the latter, changes to the SDA became a political wedge issue for election-year politicking, and it seems destined to be changed to allow for what many believe to be discriminatory scholarships. The weakening of such safeguards presents worrying signs, indicating that long-fought, hard-won protections for women are being removed.

Another eroding protection that is of concern is the Australian educational bureaucracy's declining ability to resist the change in focus from girls to boys. As mentioned in previous chapters, the bureaucracy has been a historical location of strength for "femocrats" in Australia, ultimately allowing for the initial girls' education policies discussed in chapter 2 to exist and thrive. These femocrats have also held off the most regressive portions of the "what about the boys?" backlashes of the early 1990s and 2000s, but this has become harder to do.[67] "Leapfrogging" and localizing have been the two major, interrelated techniques in accomplishing the erosion of these protections.

By "leapfrogging," a reference to the child's game, I mean that the Government has developed in BELS—and somewhat in SFB—initiatives that jump over the bureaucracy and give money directly to schools to determine its use. The bureaucracy has little decision in how the schools spend the money or what approach they take, and given the vastly easier access to simplistic and conservative boys' education books than to progressive arguments, schools are more likely to have access only to the conclusions of government-validated consultants and "backlash blockbusters."[68]

Connected with leapfrogging is the trend toward localizing programs and decision making, giving schools and communities the power to determine their own commitment to gender equity. The new draft of the recast *GEF*, again, does this repeatedly by calling for local policymaking. The government still "steers at a distance" through funding opportunities and outcomes standards, but the possibility remains for local institutions to ignore, subvert, or actively work against gender equity.[69] This open-endedness in approaches but not outcomes allows the state to deny any responsibility in creating the situation should things go wrong (what is often referred to as "exporting blame").

Controlling the Research Agenda and Availability

The purging of evidence associated with gender and education has certainly not occurred in Australia to the same extent as research in the United States under *No Child Left Behind* (*NCLB*).[70] Indeed, reports skeptical or critical of the Government's plans for boys' education are still featured on the DEST Web site. Simultaneously, though, information on girls' education has been systematically disappearing. There is, for instance, no dedicated "girls' education" page on the DEST Web site as there is for boys' education.[71] The current *GEF* (not the recast one) until recently was only accessible online from the Tasmanian Department of Education and not from more obvious and accessible sources, not even the DEST site. Other reports and policies on girls' education have languished or disappeared.

Part of the reason for the purging of gender-related research on girls is that such research is no longer being produced to the extent it formerly was. Put simply, the government is, as with *NCLB* in the United States, controlling the research agenda. Only studies that the Government finds relevant (or, politically or ideologically advantageous) get funding and get publicized. Currently, that includes only boys-related gender research.

Related to this, and somewhat counter to the U.S. situation, practitioner research (or, "action research") is currently being validated under the mandate to show "what works" in boys' education in Australia. Similar to McCracken's critiques of the politics surrounding *NCLB* in the United States, colleges of education and academics in Australia are being rhetorically assaulted over boys' education issues.[72] Attacks are frequent against academics and unions for resisting "common sense" and speaking in "jargon" that "normal" people cannot understand. In Australia, though, small-scale practice-oriented research on boys is actually winning out over large-scale experimental and statistical studies because—in the rare cases they've also been done—the latter don't always support the Government's position. Practitioner research like BELS projects, which often start with the a priori assumption that boys are suffering, have been more effective in validating Government claims. Whether such projects are sufficiently rigorous to base national policy on, though, is questionable.

Affinity Groups

Gee's examination of the central role of "identity" in education theory, research, and practice, points out that a major facet of one's identity is the "Affinity Identity," the "experiences shared in the practice of 'affinity groups.'"[73] For Gee, those in an "affinity group" may be widely dispersed geographically and may share little other than interest in an issue, but this connection can be strong. Alliances between African American parents and white conservatives to support vouchers in Milwaukee, Wisconsin (United States), provides a perfect example, for such alliances are rare and only connected to that single issue.[74] Generally, affinity groups grow organically from the interests or subjectivities of group members, but, crucially, institutions can *create* such groups to further the institution's interests. Gee gives the example of Saturn, a U.S. car company, where dealers create rituals and social events to build feelings of cohesion among customers.

This institutional creation of affinity groups can also be seen in the Australian boys' education reforms. Boys' education movements in the early 1990s began as organically created affinity groups around Australia. These drew from a diverse population—parents, men's rights groups, educators, antifeminist conservatives—but they coalesced around a "morally heated" issue and began pushing the government for policy, research, and funding. Whereas *BGIR* validated and gave voice to some such groups, the *institutional* creation of affinity groups began in earnest with the B2FM conferences. Backed by tremendous funding from the Commonwealth, these conferences created networking opportunities, "rest" days from school, and multi-sensual experiences, all surrounding a moral panic. The BELS program has created similar affinity groups, additionally providing teachers opportunities to have online chats and telephonic professional development, training sessions, and opportunities to nationally publicize their hard work. The schools get recognized as a "select" group, not only one of the few to get the grant but a model (a "lighthouse") for others to admire and follow. The key is that an institution, here the government's education apparatus, has created conditions in which diverse people can find cohesion and opportunity through a set of shared practices and concerns.

Understanding the importance of creating such affinity groups requires understanding the institutions' stake and benefit. In boys' education, such groups help the state manufacture its own legitimacy, for the state can justify its stance by pointing to the deep concerns of the grassroots groups that it has itself partly created. It also allows the state to steer at a distance by eschewing legislation; schools need only agree in advance to accept the state's position and they can, through these grants, ease (but not eliminate) the constant financial crisis they suffer. Key to the process, though, stakeholders must never know (or mention) that the state has itself created the conditions—economics, research, accountability, market pressures—through which people are coerced into "accepting" state-sponsored positions; if schools were fully funded, a grant program wouldn't be needed. Nevertheless, in the end, through such systems the state gets heroic status in the minds of voters (at least politicians hope so) for identifying a problem, being brave and farsighted enough to tackle it, and generously "giving" back to schools the funding that was really supposed to be theirs in the first place.

Positives

I don't want to leave this chapter having given readers only the impression that the government initiatives in boys' education are all bad and that those who have enthusiastically participated are somehow duped or suffering from false consciousness. Indeed, there are numerous potentially positive implications to these programs.

First, and perhaps most important, the initiatives following from *BGIR* have created much practice-oriented knowledge on boys and pedagogy. Some of the BELS projects, particularly, have filled gaps in our understanding of what does and does not work for particular boys in school and why. As Lingard and colleagues point out, often such trials lead schools to abandon regressive or stereotypical practices in favor of balanced, gender equitable pedagogy.[75] Also, many of the BELS programs have put technology to good use, have focused on disadvantaged boys, and, frequently, have also reported helping girls.

I must also note the commendable amount of autonomy and professional respect given to school-level educators through these various programs, particularly BELS. Despite whatever potential for "leapfrogging" and policy control through localization, the action research and teacher empowerment demanded by the BELS programs have been tremendously effective, providing many teachers opportunities to learn highly transferable research and pedagogical strategies. I have elsewhere argued that such skills are key to overcoming the simplistic, regressive, stereotypical "solutions" for boys' problems that teachers encounter in popular press books and the media.[76] A concrete example of a group of teachers doing just that is presented in the next chapter.

Implications for the United States and Elsewhere

Though the United States and other countries have not seen concerted boys' education policy to the extent of Australia's, many of the same dynamics as those evident in the resulting initiatives to *BGIR* are present outside Australia. It does not take national level policy to create an economy of boys' education (though it clearly helps).

All over the world, conferences, programs, materials, and consultants have developed a web of financial incentive to regard boys as endangered. A *New York Times* article, for example, details the rise of consultants to help boys get organized, a skill "what about the boys?" advocates lament that boys often lack.[77] Conservative Christian groups, too, as I have shown elsewhere, have attempted to capture this discourse to sell toys and costumes.[78] In addition, of course, consultants and popular press authors have made money in the United States and other countries, as much as or more than the amount in Australia. Michael Gurian's institute's Web site sells newsletter subscriptions at a rate of US$9.90 per issue, and it lists US$40 parenting videos, all of his books on sale, and contact information for on-site training and their three-day, US$645 summer institute.[79] They'll also train people to become trainers for the Institute for US$2000.[80] And even without U.S. federal funds earmarked especially for boys' education, the Gurian Institute has a document that outlines how schools can gain grants under *NCLB* to pay for his consultancy.[81]

Affinity groups, too, have grown up around boys' education in other countries, often through the means of conferences and professional training, as it happened for the B2FM conferences and the BELS program. Groups in the United Kingdom and the United States have coalesced around boys' issues, and educators of every political stripe have been welcomed into the fold. A recent experience of mine keynoting a local Writing Project workshop, for instance, saw teachers from all over the upper North Dakota and Minnesota region communing with other likeminded teachers about the difficulties of getting their boys to read and write.[82] The inclusion of a professional book and lunch made even this Saturday morning event a big draw. Gurian's professional development groups provide similar opportunities.

Boys' education has also been used for scoring political points in other contexts than Australia. For conservative politicians all over the world, boys' education—partly because it has certain kinds of data to back it up—has been a discourse useful for playing to the antifeminist sentiments and commonsense understandings of their constituents. In Canada, for instance, public antifeminist discourse has been prevalent and growing, and such discourses have led to, among other recommendations, hiring more male teachers.[83] Similar legislation for hiring more male teachers has occurred in Iceland as a reaction to boy debates, and U.S. groups are increasingly sounding alarms to find more men for classrooms.[84] Politicians in the United States, too, are scoring political points on boys' education beyond calls for more male teachers, a point I expand on in chapter 6. Clearly, the dynamics underpinning interventions created for boys' education have global relevance.

Conclusion

In this chapter, I have largely taken a theoretical and descriptive—but often grounded, too—view of the initiatives following from *BGIR*. Viewing the B2FM conferences, the male teacher strategies, the *GEF* recasting, and the BELS and SFB programs from this vantage aids in understanding the macro-sociological and political import of these schemes. I will move in closer to the boys' education ecology, though, to truly understand how such policies and programs interface with actual schools and those who run them and teach in them. In the next chapter, then, I focus on what has happened "on the ground" in some schools that are struggling to reach their boys.

Chapter Five
"Getting It Right" in the Schoolhouse: Two Case Studies

Teachers are trying to figure out things on their own [to engage boys], in a school setting, because the states have not really taken this up.
—Richard Fletcher, *Hansard,* 1054

In this book, I started with a satellite view of the global dynamics of boys' education, moved in to focus on Australia, and now I change instruments, using a microscope to observe the closest I will get: the micro-ecologies of specific schools and classrooms that are addressing boys' education "on the ground." This chapter looks at two case studies of schools playing out the day-to-day struggles to "get it right" for boys, as *Boys: Getting It Right* (*BGIR*) would phrase it. One case examines a cluster of public or "state" schools awarded a Boys' Education Lighthouse Schools (BELS) grant. The other investigates a private religious school that has almost complete policy freedom.

Importantly, these case studies are just that: cases. No case study—including these—can represent all the various reactions to a policy. Complete representation, though, isn't the point. Sheridan, Street, and Bloome stress this in their distinction between a "typical" case and a "telling" case.[1] I am not endeavoring to show the *typical* manner in which schools take up boys' education and the *BGIR* report. Rather, these cases are "telling" instances, ones that demonstrate just some ways these issues impact real schools.[2]

For all that these cases tell us, however, they cannot answer every question. While, for example, I can gauge how *BGIR* influenced these schools, I cannot gauge the diffusion of the report to the entire country. Also, while I can discuss how administration and department heads at the schools structured programs, I cannot address whether *BGIR* influenced students' and teachers' day-to-day interactions. My focus was on the planning and decision-making occurring before or outside of the classroom. Even with these limitations, though, there was much to be learned about the boys' education policy ecology in Australia from the total of eight schools I researched.

I begin this chapter by presenting the case of the Riverside Schools Cooperative (all teacher, student, and school names are pseudonyms), a group of seven schools awarded a BELS grant during that program's first phase. Their case demonstrates the power of money (or its absence) to shape and destroy programs regardless of participants' interests. It also highlights the progressive potentials within the boys' education movement given the correct strategy, for the program was reframed early to use a "which boys?" perspective. The second case describes the boys' education work of Springtown Religious School (SRS), a private, coeducational school with independence from both governmental and denominational policy. Such independence does not mean complete freedom, however, for heightened concern for school culture, tradition, and a particular educational philosophy led to constraints of their own. To end the chapter, I explore the lessons that these cases teach us when considered together, particularly about why educators sometimes take up boys' education, why they sometimes never take it up or drop it, and how schools utilize policies like *BGIR* and other information sources.

The Riverside Schools Cooperative

The Riverside Schools Cooperative (RSC) is a group including one secondary school (Gretna Green State High School), and six primary schools that are the high school's "feeders" (Anchor Creek, Point Lookout, Sunshine Valley, Marntal, Hyde, and Summit State Schools). All are state (that is, public) schools. How a school—or a group of schools—educates its students always depends to some degree on the context in which it resides. The RSC is certainly no exception, for its context has changed significantly, and the RSC has had to change with it.

Context of the RSC

The city where Riverside operates is geographically one of the largest in the Southern Hemisphere, but it centers on a river where the hub of culture and business lies. Beyond this stretch along the river lie mainly residential areas dotted by patches of commercial and retail structures. This was for most of the city's history an agricultural area, providing fertile soil for cotton, sugarcane, maize, wheat, and bananas. With industrialization it became a working-class section of the city, where factory labor and middle management for industry lived, worked, and shopped. Recent history has seen a relatively large unemployment rate in the area, but this has changed drastically in the last few years. Much is still changing, in fact. A particularly trendy street, lined with eclectic shops and popular restaurants, has attracted more cosmopolitan, professional residents. Still, for all the historical difference, the area's current demographics makes it relatively average when compared to the city as a whole; the RSC's neighborhood makes a prototypical case, statistically.

While in some places the altered residential patterns resulting from gentrification might revitalize the public schools, the Australian dual system of government schooling and private schooling has made these changes complex for the RSC. Many of the new professional families are young and without children, but many, though clearly not all, with children have followed the general trend toward private schooling popular with the "aspirational middle-class" (a popular phrase for this class fraction during the 2004 election). Government schools have seen large dips in enrollment as private schools "creamed off" the more affluent and able students, leaving behind students slightly more likely to have learning or behavioral difficulties and more difficult home lives—the kinds of difficulties private schools often cannot or will not address. With private schools' additional advantage of government funding in addition to tuition, many state-run schools have found it hard to compete in terms of facilities, programs, and opportunities.

Compete schools must, however. As enrollments declined, the very survival of schools in the area looked to be in question. Even retaining teachers, despite usually better salaries and benefits in state schools, is difficult compared to private schools' sometimes better labor conditions. Government schools have been forced to constantly reform and develop niches if they are to compete with private schools.

Vast differences in important community factors, moreover, make certain that RSC schools are even more vulnerable. Most of the schools are small (especially by U.S. standards), but they vary widely in size. Anchor Creek had 475 students in 2003, while Summit had 115. Class sizes varied from nine students for one high school course to 33 for a primary school class. The communities from which these schools draw vary highly as well. Marntal, Point Lookout, and Sunshine Valley are the most disadvantaged communities and schools in the Cooperative. They have more non-English speakers, Indigenous residents, single-parent families, low-income earners, and unemployed persons. At the same time, they have fewer median incomes, high-income earners, holders of college degrees, car owners, and home computer and Internet users. It should be of little surprise that, given these characteristics, fewer residents of these areas send their children to private schools, particularly at the secondary level. This also means that public schools, which are generally less well funded than private schools, must expend more resources coping with these needs. In turn, the thinner-spread resources fuel the exodus of middle-class children, whose parents are seeking educational advantages for them. It is a vicious spiral.

The RSC was formed from these schools that were previously only nominally connected, in part as a way of combating the challenges of fewer resources and increased competition. As the Cooperative's promotional literature (a sign of their need for marketing) recalls:

> Over two days in January 2002, staff from a cluster of seven schools met to participate in the process of establishing a network to work [together]. So began the [RSC]. These schools are working smarter, capitalising on the expertise of the larger number of teachers, building effective partnerships with parents, the community and business—innovating, sharing ideas and resources, rationalising planning, developing curriculum and staging events.

The language of business clearly shows in this mission statement—"innovating," "capitalizing," "expertise," "effective"—reflecting both the competition impinging upon the schools and the turn toward business models as a solution to this competition. Similar discursive methods have been used in the higher education sector for years, and demonstrate the reality of spending time and resources—both scarce—for the marketing and commodification of public schooling.[3]

The RSC's sharing and resource consolidation can be seen as a reasonable, even progressive, response to hard times in funding. These schools are getting teachers to talk to one another, share ideas, and cooperate in putting together programs and activities that otherwise would be prohibitive in costs and labor if done alone by one school. It also allows for the multiplication of children's social spheres, as they get to meet many more students from their area. This gives the RSC resources of diversity that some of the schools wouldn't otherwise have.

The BELS project the RSC undertook in 2003 encapsulates well its mission of cooperation and resource sharing and was one of their first major cooperative efforts. Though spearheaded by a small group from Gretna Green High School, the project was conceived early on as one in which all of the schools would participate. Equal numbers of teachers would be involved from each school, and the thrust was always to solve a perceived common problem: boys' literacy difficulties. Cooperation, further, made sense for these schools, both economically and discursively, for the grant allowed a single school to apply for A$5000 (US$4363) while up to A$20,000 (US$17,450) was available for groups of four or more schools.

Beginnings and Feared Endings

When I first went to Gretna Green High, the main site of the activities of the grant, Mr. Philips met me and led me to the conference room where we wouldn't be disturbed by phones and other distractions. A trim man, he was wearing an open collar with a thin, tan jacket; he was not tall, and he had close-cropped beard and hair, both reddish in color but beginning to show gray. He rarely looked at me as he spoke. Giving me a copy of their grant proposal from within a thick folder of materials on the project, he explained that they were trying to create a project that would tackle boys' literacy skills, especially writing, as well as boys' motivation and resilience. The main features, he explained, would be action research in which teachers would find out their boys' learning styles and a role model project that would have boys come into contact with male authors during the RSC's upcoming literacy festival.

After explaining their project, he then asked how I thought I might be able to fit in. I told him I saw my role as best suited to evaluation and data collection. He seemed pleased and a little relieved, saying that my taking those roles would free up the project directors for other work. He invited me to a meeting of participating teachers (about 12 to 14 in all, one or two from each of the seven schools) the next week at an off-campus site. This meeting did not take place, though, and I could not get in contact with Mr. Philips. I was beginning to get worried that the project would not happen.

Unbeknownst to me at the time, soon after this first meeting, when I was beginning to feel that it had been too long since hearing from the school, perhaps the most shaping event of the entire project had occurred: Mr. Philips had taken on a new principalship elsewhere and a new principal, Mr. Garfield, had just been appointed. With BELS report deadlines looming and the threat of losing the grant money clear and present, the new principal had, with some trepidation, agreed to take on the project. Much time had elapsed on an already tight timeline for the BELS program, though, and major modifications would have to be made. Now, rather than taking three months, the RSC would have to complete its project in about three weeks. Activities would have to be rearranged, teachers would have to make quick scheduling accommodations, and participants would have to do quick research.

The Project

The major events of the original project that addressed teacher professional development and research changed little with the project's shortening. The main focus was still action research on boys' literacy, and authors of children's books were still going to be called in, though now only the teachers would interact with the authors (the hopes of having students interact with them had to be put off). So the schedule of events was planned as follows:

1. Martin Mills, an expert on boys' education from a nearby university, would come and give a half-day workshop on action research and help the fourteen teachers, two from each school, come up with a plan and research questions.
2. Teachers would collect data in their classrooms on their research questions.
3. One children's author would give a presentation after school for the entire RSC on boys' literacy.
4. A poet and author of children's and teen's books would give only the project teachers a full-day workshop on engaging boys with popular culture; another famous children's author and university lecturer would give another full-day hands-on workshop on developing boys' literacy skills.
5. Mills would again have a half-day action research workshop, this time focusing on the results of the teachers' research.
6. I would interview the teachers about the project and collect their data and artifacts to eventually write a final grant report. (Though I was concerned about blurring the lines between participant and researcher by accepting such a responsibility, this level of participation gave me special status, for I was doing something for the group that seemed to them difficult and time consuming, and so some saw me as helpful, self-sacrificing, and even "brave." In return, I had a reason to be ever-present.)

From the interviews with teachers, clearly the action research process was the part they approached with most trepidation. This was the first activity, and little had been explained to them in advance. With the change of principals and some problems with the E-mail system, communication had been difficult and sparse. As Ms. Elliott from Anchor Creek School put it, "I kept going to [the principal] and saying 'When is it happening? What do we have to do?' And he didn't know."

Still, whatever unease there was in coming into the action research, these sessions were to be a primary shaper of the project's direction. In his first contact with the teachers, Mills clearly laid out a (pro)feminist perspective. Discussing the findings of his recent government-sponsored work on schools with programs on boys' education, he said:

> One school we went to was a Queensland high school that *had* had a boys' education program, and they dropped it because what they saw was a whole range of scrapbook changes that were making no difference to the boys' outcomes. And what they went to was a focus on pedagogy and a focus on doing things in the classroom, and that was making a difference in the boys' outcomes. They were saying, "Good pedagogy for both boys and girls makes the difference."... Some of the common features of those schools [that were successful with their programs] were that they really had done thorough research into the kids at their school. They were aware of the "Which boys? Which girls?" approach—that we *cannot* talk about boys as a whole group across the state. The boys on Thursday Island are different from the boys at [Gretna Green]... to some extent—different pressures on them. And having to understand issues of [boys'] gender *and* class, ethnicity, race, all those issues are very, very clear.... *I* would imagine there are boys in your schools that are doing fine, right, and their literacy levels are fine, but that there are boys that are *not* fine. You would also have girls that are possibly not fine in terms of literacy. And there is a big danger of treating all boys the same or all girls the same *or* of making assumptions that all boys are failing, all girls are doing well.

He then emphasized the point by having teachers think about the different *kinds* of boys that go to their schools, drawing on the typology work of other researchers—he mentioned R.W. Connell and Wayne Martino as examples.[4]

This early framing by Mills of the debate as "which boys?" set a theoretical perspective that would be remarkably pervasive amongst the teachers and would guide the direction of the project. As one example, Mrs. Hirst (Gretna Green) explained,

> I guess my initial interest was probably protecting girls' interest because I was worried about the direction it [the project] was going to take, but my earliest experience of professional development was along the lines of Martin Mills, "which girls and which boys?," and that's still in some way my philosophy, that it's more about socioeconomic status and that's certainly borne out in our research here.

Here Mrs. Hirst uses the exact language of Mills' (pro)feminist critique, suggesting that his perspective was influential. This isn't to say that some didn't already hold this perspective, but given the dominance of simplistic and conservative understandings about boys—that schools and literacy don't "fit" them, that *all* boys are troubled—having this (pro) feminist perspective hold sway *ubiquitously* across a group of teachers should be instructive.

The ultimate result of the action research workshop, then, was to formulate a group project that all of the fourteen teachers, regardless of their grade level or interests, would work toward. By the end of that first session, the teachers developed (though some apparently only went along with) two basic sets of questions: (a) How do teachers (or how can teachers) engage boys with writing? Why do some boys engage in writing? (b) What causes boys to change as they get older? What are these changes—academically, socially? It was also determined that, by the next meeting, each school's teachers would bring a description of their school, identify students to focus their research on, and share some pilot data obtained.

In the intervening 19 days, the participating teachers took part in sessions with three authors whose explicit focus was to be on issues of boys' literacy. The first explicitly focused on engaging boys in reading rather than on teaching methods. The speaker pointed out several statistics that suggested boys were doing more poorly than girls on reading tests, but—in the manner of asking "which boys?"—he noted that upper class boys were doing well compared to their working class peers. He also proposed to the teachers a "chicken and egg problem" of which comes first: poor literacy skills or reluctance to read? In general, he focused on the socialization of reading and the social world's impact on whether boys read.

The other two authors conducted back-to-back all-day sessions for the 14 teachers involved in the BELS project. The second of the three authors was a poet and author of novels for young adults. He spoke with passion about using poetry and historical nonfiction to engage boys in writing, and he walked the teachers through specific exercises that could be used in a classroom. The focus on boys was not done uncritically, though. Indeed, he said early in the session, "I'm not entirely certain it [engagement in literacy] is a gender issue." There are, he told the teachers, huge "vacillations," "swings" in education that often end up "throwing the baby out with the bathwater." What was learned in the work for girls, in other words, was being overlooked or tossed aside in the rush to boys' issues. Nevertheless, frequently punctuating the activities he demonstrated were mentions of what to do "especially for boys," to gain the boys' interests. This included building from their own experience and that "boys like stuff they can control, stuff that is way out there."

The third author was a celebrated children's writer. He framed boys' literacy using research that noted boys perceive a disconnection with the curriculum.[5] He also brought up with the teachers secondary sources that address the biological dimensions of the issue, particularly work on brain differences, but mainly he impressed upon teachers the notion of gender as constructed by showing historically changing conceptions of what characteristics were valued in boys.[6] Saying that he came to the issue because of his own son's reluctance to read, he promised to give the teachers "practical exercises" that he said are "guaranteed not to fail." This included using an ambiguous picture as a writing prompt; using and modeling the "plot triangle" that demonstrates exposition, climax, and resolution in a story; using maps of unknown lands to prompt fantasy writings in the vein of J.R.R. Tolkein; and guiding students through close readings of children's books to show them how to identify motifs and use illustrations as context clues. The teachers were taken with the passion that this author conveyed, and most reported feeling energized when they left after these two days, many of them excited about implementing some of the activities in their own classrooms right away. For some of the teachers, this wish to begin implementation of the strategies had to wait, though. It was the end of term, a time of grading and reporting, and yet another pressure on them that took away their attention, according to them, from the BELS project.

It was at this point, just after the author workshops, that I began traveling to each of the RSC schools to conduct interviews with the teachers. For each interview I asked the teachers about their participation in the project, their interest and experience in boys' education, and what they did for their action research project that would contribute to the final report, which I was to write in the coming days.

What emerged from the teachers and Mr. Garfield, the project coordinator, was a notion of "local solutions to local problems"; this had been suggested by Mr. Garfield in an interview, and it tied together the action research component and the participating teachers' oft-repeated notion that the focus shouldn't be on simple solutions given as if all boys were in trouble, but that instead the focus should be on the specific problems of the boys in a specific classroom.

The findings from the teachers' research on their own students' literacy demonstrate clearly the limitations of popular arguments about boys. The teachers found, sometimes to their surprise and clearly contrary to the circulating stereotypes, that, among other things:

- Most of the primary students like to write, even those who are reluctant.
- Significant numbers of girls are also disengaged or in need of help in writing.
- There is more parental support for their boys' literacy than expected and significant numbers of boys have positive male role models of literacy, both at home and school.

- In some schools, being a good writer was not stigmatizing, while in others some boys disavowed academic abilities.
- A larger than expected amount of literacy sharing occurred within boys' peer groups.
- For some it was harder to get girls started on writing, and girls' writing was often stiff, formulaic, and lacking interest compared with boys' risk-taking with plot.
- Many boys are reluctant to revise their writing, preferring to finish and get on with the next thing.
- Many boys write only when they must, not for pleasure.
- Students' definitions of "writing" or "good writing" often conflict with teachers' definitions—students often equated "writing" with handwriting, neatness, or mechanics.

Of course, no finding was universal and all had contradicting evidence. Still, these were generally agreed to be important during the second action research workshop when the teachers came together to compare what they had found. Several expressed an interest in pursuing the topics and strategies implied by these findings, hopefully, they said, with more time and a more long-term focus. All, though, clearly felt their thinking or practice had changed—or at least that it *would*—having participated in the project.

After the RSC report was done and submitted, there was, naturally, a waiting period wherein the BELS program administrators would process the data submitted by the more than 200 schools, which eventually became the report, *Meeting the Challenge*.[7] It summarized the findings of the RSC's project—and all other projects—but little other contact occurred between BELS administrators and the RSC. No formal evaluation of the RSC's project was made by the granting agency; it was all self-report (a fact that underscores the precariousness of using BELS findings as a basis for national policy).

Soon after, the application period for the second stage of BELS began, and the RSC was eager to get selected, particularly since the funds had grown to A$100,000 (US$87,250) for each school or cluster picked. They again asked me to help with their application, this time via E-mail from overseas. In the new proposal, the RSC, led again by teachers from Gretna Green, proposed extending the action research project from the first stage, this time with more training from Martin Mills (seven sessions), full teacher release funding so that more teachers could participate, and with "consumables" included in the budget (last time this had been neglected and it was difficult to find A$50 [US$44] to buy videotapes to finish the project). The application also requested funding for an RSC Web site on boys' education and further funding for the annual literacy festival. Finally, they included workshops for boys with a group that teaches students to do circus performances as a means for building self-esteem. Though a detailed and specific plan, the RSC was ultimately turned down for the second phase of BELS. At that point, their *systematic* plans for focusing explicitly on boys' education withered and disappeared.

What We Can Learn from this Case

Several important issues arise from the RSC's BELS project, how the teachers conducted it and their beliefs about it. First, it clearly demonstrates the progressive potential of action research to combat simplistic, regressive, stereotypical, conservative, and antifeminist notions that often circulate within the boys' education debate. Though the brief projects didn't develop much new pedagogy, teachers clearly had their (some might say deficit) notions challenged by just talking to their students. Teachers talked of being "shocked" and "surprised" by findings that boys *did* have male role models for reading, that their boys liked writing and reading, and that the boys had passion and skill with language. Also, the teachers learned concretely through their research that it isn't *all boys* that have difficulties, and thus all expressed realizations that they had to constantly ask the common (pro)feminist retort to "what about the boys?": "Which boys?"

Second, and related, the theoretical groundwork laid was important to the project's tone and direction. Mills articulated from the beginning a progressive, (pro)feminist stance on the education of boys, one drawing on common sense and experience in teaching, and the teachers all took it up. While accepting this stance might not be solely a result of the project, these teachers' ubiquitous determination to ask "which boys?"—in opposition to the most popular and available literature, to the prevailing sentiments in the government, and even to much of the RSC's initial project proposal—suggests that Mills's (pro)feminist message was likely significant in shaping their approach.

The RSC case also raises questions about the sex of adults who participate in boys' education reforms. In the RSC project, it fell disproportionately to males, for four of the fourteen teachers were males, a higher proportion than the proportion of male teachers in the RSC. While certainly having men take responsibility in educating boys is good—reversing the trend to heavy emotional labor for women in the restructuring of schooling[8]—such expectations could be extra discouragement to males becoming teachers, as the toughest cases and lowest-performing students may be heaped on them. Several of the males in the RSC's

BELS project describe feeling forced into the project simply because they are male. Mr. Shay from Marntal State School, when asked how he got involved, for example, said it was from

> Sort of being pushed in that direction, I guess, from the boss [his principal]. So when asked for volunteers, someone had to do it, so I thought, "Well, being the only male apart from the principal"—the only classroom teacher such as it were, the six of us—I thought, "well, I better put my hand up."

Many of the female teachers also felt forced into the project, but largely this seemed to be a function of small staff sizes and needing two participants per school. No female teacher claimed that she was picked *because of* her gender. Again, more responsibility for boys' education accruing to males isn't inherently or necessarily a bad thing, but the effectiveness of programs where the participants feel coerced or lack requisite skills and threshold knowledges is questionable.

The RSC example also demonstrates the power of money and competition to shape school reform. Not only did the money involved determine what was possible and how many people could participate, but the money's presence was also a cause for the project originally. While the Gretna Green teachers and administrators who applied for the grant had some concern over boys' education already, their ability to address it would have been more modest without funding. In addition, keeping programs solvent invariably requires that schools chase grants and market themselves as having a niche, like boys' education, to attract more parents who can pay voluntary school fees and otherwise contribute to the school.[9] To some extent, the focus of boys' education was part of the RSC's marketing strategy, for the BELS grant was used in newsletters, on the Web site, and in brochures as one example of their success. As schooling in Australia moves closer to privatization, with the large sums given to private, fee-based schools, and as education moves more toward site-based management schemes, such business-minded practices may increasingly drive the educational decisions of principals and policymakers. This mindset also drives concerns about boys' education.

Springtown Religious School (SRS)

Springtown is in many ways the other side of the coin when compared to the RSC schools. Being private and religiously affiliated, it is the competition. Still, the differences between these two cases are key to understanding the diversity of responses to boys' education policy and what difference in context brings about. Springtown's example provides several approaches taken to boys' education from across the school's curriculum; the RSC's project was somewhat contained by comparison.

SRS lies north of the river. In comparison with the average neighborhood of RSC schools, on many indicators the community surrounding Springtown is doing similarly or, in some cases, worse. To a great extent, though, community demographics don't reflect the school population, for, like most Australian private schools, Springtown draws students from a wide geographical region, not just its neighborhood. Some students travel nearly an hour each way on public transportation to attend. They're enticed by the school's reputation and uniqueness.

Springtown sits on an attractive campus—more polished than its neighborhood—with manicured lawns, umbrella trees, brick benches, and low buildings, all clean and with immaculate landscaping. A whirlwind of activity is assured between classes. During morning teatime, students loiter on the front lawn and in the spaces between buildings. The primary-colored student uniforms stand out in sharp contrast against the greenery and—depending on the season—the various oranges, violets, and pale blues of the flowering garden beds. The boys and girls arrange themselves in almost solely gender-segregated groups. Groups of boys hang around the benches while girls huddle in almost perfect oval groupings on the ground, most blocking the sidewalks. There is a large, state-of-the-art performing arts center, several classroom buildings, a gymnasium, and a library. Each is connected by a series of twisting walkways and short sets of steps, and interspersed between buildings are garden beds, sculptures, and seating areas, most of them shaded to protect students from the harsh northern Australian sun. The school also houses a uniform and stationery shop, a tuck shop (what Australians call a canteen), and a daycare open to the community and partly staffed by students. At the back of the campus are sporting courts and a sports oval (the appropriate shape for both cricket and—less common in the north—Australian Rules football).

The school's population, both students and faculty, reflects a largely white, middle-class, Anglo-Australian background. The school is coeducational and, in 2003, had a roughly gender-balanced population of about 555 students in Years 8 through 12. The administrators, who called themselves the "leadership team," made a point of using the school admissions process to "keep it as close to [50/50 boys to girls] as we can." There were 55 faculty members (about two-thirds women), most of them teaching full-time. The leadership team consisted of: (a) the principal, Mr. Spears, a bespectacled, graying man with a frame befitting his former rugby playing days but a gentleness and quiet demeanor more characteristic of his supportive leadership and his big-band piano playing; two deputy principals, (b) Mr. Powers, a stylish, inquisitive man who was always quick with humorous Australianisms, and (c) Mrs. Howard, a warm, enthusiastic woman with shoulder length blonde hair and an engaging affect (both deputies also taught classes, in math and English, respectively); and (d) the business manager, Mr. Lockyer, a blonde, cleancut man with a wealth of advice for good travel spots. There was also support staff, including secretaries, office personnel, and grounds staff. Staff and faculty also did many of the ancillary duties of the school—tending to sick children, counseling, and so on—whether formally or informally, as needs arose.

SRS classifies as a "moderate fee school," with a 2003 tuition fee of A$4800 (US$4188) per year. According to the school prospectus, several other "typical costs of a year" are involved, though; so for a family sending, say, one son for Year 8, letting him learn the tuba and play basketball, they could expect to pay A$6843 (US$5971) in a year. Considering the citywide median annual household income of A$36,400 to A$51,948 (US$31,759 to US$45,325) as reported by the city council, that tuition represents between 13.2 and 18.7 percent of the average family's income. Most but not all of Springtown's students, however, come from slightly higher socioeconomic groups.

Private schools in Australia, unlike in the United States, are, again, partly funded by the Commonwealth government, so these fees don't represent the total resources of the school. Rather, the school's finances, according to Mr. Spears, the principal, "have been carefully structured to balance government funding with tuition fees." Government funds account for about half the cost for each student, a 50–50 split the school achieves partly through staking a middle ground in fees.

Despite government funding, Springtown, like other private schools, has a great degree of policy independence. It isn't subject to most state or federal educational policies, though it does have to comply with basic rules of teacher qualification, facilities, and academic performance to qualify for accreditation and, thus, federal funds. Additionally, the school has been granted leave from its affiliated church to do its own policymaking, without having to follow the strictures that schools of the same denomination do. This policy independence from the state and from its religious denomination makes this case particularly interesting and important, for it illuminates what teachers and administrators do or don't do with policy when coercion and mandate have been removed. It also gives insight into the role of professional discretion in boys' education, and the use of de facto policy in the absence of de jure policy, as I discussed in the Introduction. The question thus becomes, from *where* do they draw policy, not *if,* for all schools—to differing degrees of formality, prescriptiveness, and success—must establish some internal policies. To say that the school is policy independent, though, does not mean that the teachers within it have complete freedom to do what they want. Deep concern for school culture, tradition, and a particular educational philosophy led to constraints of their own.

Indeed, a major shaping force of nearly all activity at Springtown—boys' education being no exception—was the pervasive notion of Springtown's uniqueness, it's idealized identity, and the preservation of the "school culture" that had been established at the school's founding in the early 1980s. As Mr. Spears, one of those founders, explained, firstly, they eschew the single-sex tradition of other private religious schools, a controversial move, even today, when separate spheres has gained new cachet in conservative "what about the boys?" discourses. Second, they strive for a "cooperative and collaborative ethos" rather than competition, largely through shared, distributed leadership between faculty and administration. Perhaps the clearest expression of this ethos comes in the noncompetitive foci in curriculum, particularly having outdoor education (like ropes courses, kayaking, and hiking) rather than competitive, interschool sports (like rugby and netball).

Another key and readily apparent feature of the school culture, particularly the faculty culture, is a strong commitment to professional development. The administrators take sabbaticals every six or seven years; all had been to Harvard University to study leadership, and each had done other projects around Australia and overseas. Classroom teachers, similarly, were often funded to go to national conferences, including to the 2003 Boys to Fine Men conference described in chapter 4. A list made in 2000, in fact, included 42 professional development events attended by Springtown faculty and administrators that year.

Each person or group funded for sabbatical or a conference was expected to share what was learned in some way. The school semiannually publishes a professionally printed, full-color magazine including school news and teacher reports on their research, professional development, or experimentation. Also, Springtown annually puts together its own *Professional Journal,* a bound, photocopied collection of numerous reports from conferences, teachers' reflective essays, program overviews, and occasional reprints of short articles. This continuous sharing and reflection forms the backbone of the school's professional life.

A surprisingly muted aspect of the culture was the school's religious orientation. I never directly asked about it, but the subject also rarely came up spontaneously in interviews, even when talking about the school's core aspects. Some students evinced a dismissive view of the religious curriculum, and several parents, according to research by an independent evaluator, felt the school was not focusing enough on religion. Still, despite its quiet place in my research, the religious orientation was clearly a key element of students' experiences. There was a palpable tension between following gender maxims of its relatively conservative Christian denomination and the progressive values of equity and cooperation central to the school's mission.

The school culture at Springtown—its uniqueness, its exceptionality, and its purity to the founding vision—was closely guarded by the leadership, and nearly every decision was weighed against its impact on this culture. For the leadership, there was constant pressure from within and without to move away from the founding principles, a move that they perceived would destroy the school's exceptionality and unique draw. These principles had been hard-won in the school's founding, and they were constantly, explicitly discussed within the school community. Disruptions and new community or cultural concerns were often viewed as threats to this carefully maintained ethos. As administrator-produced handouts from a staff meeting about the culture stated it, "We need to be proactive in shaping our school culture in a turbulent and sometimes threatening environment, or it will be shaped for us by others."

Boys' education issues clearly were a new, potential threat to the culture. Consider the defining moment, the "impetus," for Deputy Principal Powers that convinced him that boys' issues needed attention.

> Mr. Tully got up and said essentially two-thirds of the student management issues related to boys.... [Tully asked,] 'What are we going to do about that? Rather than just react to it all the time, is there something we can do to stop that?' And so that was, for me, a really big incentive to think.... He'd done the statistics for here, and that's what surprised me, 'cause *I would have thought here was a bit different to other places*. (emphasis mine)

The "surprise" of the newly discovered problem for boys troubled his notion of Springtown's exceptionality and the culture that creates it. The solution, then, was for the school to intervene to correct this imbalance. Beliefs in one's exceptionality can often (but not always) suppress the ability or willingness to see problems. Once seen, though, any problem has to be checked against the need for maintaining the school's ethos as originally envisioned.

This heightened vigilance for threats to school culture had a significant, determinative impact on the advancement of boys' education concerns. Such gender concerns presented a challenge to their core beliefs in coeducation and equity. To pay special attention to boys, in other words, was to violate key principles. Consider Mr. Spear's declaration during our first meeting to discuss the research:

> We really do believe that, especially during adolescence, that coed is the way for boys and girls to learn. [*inaudible*] because that's the way the world really is. But we haven't taken on a lot of structural approaches to things such as boys' education [*inaudible*] of schooling, but we've really gone for the cultural elements to try to build it into school culture all the things that seem to us to be important.

Their approach, instead of being "structural," could be summarized in the way Mr. Spears describes an excursion organized by Mr. Tully, who was informally charged with dealing with the school's disengaged boys:

> [Mr. Tully's] first initiative after [noting that boys were disciplined most] was to organize a father-son fishing weekend, and we had that. It was good. I mean they are little things that have to be part of the broader framework. But you know those little things, when you weave them all together into the culture of the school, can be really helpful. That's the first thing I can recall that we've ever done on a single-sex basis in this school. And of course some of the girls were a bit annoyed. They like fishing, too. So he's now on a promise that he's going to take the girls fishing.

This last reaction, to quickly point out the violation of the coeducational ethos, normally maintained by carefully balancing all programs and events by sex, was a common one. Attending to boys' needs was an apparent threat to the school's core philosophy. Still, the pressure was there, from both within the school (from key leaders in the faculty) and from without (including parents, *BGIR*, and the media barrage), to focus on boys. Thus, an accord was struck in the general approach for Springtown in boys' education: "little things" woven "all together into the culture of the school."[10] One of the school's major approaches to reform for boys' education, however, violated this general approach.

Millennial Kids
The ~~Boys' Education~~ Group

The school counselor and coordinator of the Personal and Spiritual Development (PSD) curriculum was Mrs. Sanderson, a small, energetic woman with an upbeat personality. When, as I mentioned before, Mr. Tully—a reserved man with earrings and long dark hair, usually in a loose ponytail—came to the Student Management Team (SMT), the discipline group, to announce that boys were disproportionately disciplined and that proactive measures needed to be taken, Mrs. Sanderson was enthusiastic to help. They decided to fix these problems by starting a Boys Working Committee, which also included the two deputy principals and three other teachers. Mrs. Sanderson and Mr. Tully were funded by the school to go to the 2003 Boys to Fine Men (B2FM) conference, and they did much research on ways to "reach" boys who were the "hard nuts."

One source the two looked to for information was *BGIR*. I asked Mrs. Sanderson what she thought of it.

> Well, my first thought was "Well, what is here that's new?" [*laughs*].... My second thought was I went through and highlighted and chose nine things that I thought Springtown could do.... I got the Inquiry [report] early at the start of the year, which is when I said to Mr. Tully "Come on, Mr. Tully, let's go to Newcastle [to the B2FM conference]."... So yeah, I've read them [the recommendations of *BGIR*]. I'd say that out of the 24 recommendations... there's only nine that I thought were areas that we could work on.... One of the things that was in there that Mr. Tully and I support is single-sex classes for English and science. But it [our suggestion] was knocked flat. [*laughs*]... So, yeah, recommendations [of *BGIR*] are good, but it's like a government policy, and you have to sit there and go "Well, if we put let's say the nine that I thought we could do here, where do we get the *money* to implement these things, you know?"

Clearly, some specific ideas for programs came directly from *BGIR*. The other major source of information for the Boys Working Committee was a series of focus groups the teachers conducted with single-sex groups in Years 9 and 11. The girls and boys both, Mrs. Sanderson told me, said similar things, including that boys were treated more harshly in discipline. Considering that's also where the movement toward boys' education began at SRS, with discipline records to prove it, it is safe to say that the problem was genuine.

The other major issue that concerned the new boys group, Mr. Powers said, was the results for Year 12 students, especially how boys compared to girls. Based almost solely on Overall Placement (OP) results, a measure used for university entrance, they concluded that both sexes achieve similarly; the boys were *not* underachieving academically. Mr. Powers noted that this isn't the same as many other schools, where girls and boys have divergent OP results, with boys behind. Springtown's results have been consistent over the last ten years, since they began keeping the statistics.

Their main data for grades 8 through 11 are Chairperson of School Council Awards, given to students keeping a certain number of A's in a certain number of subjects. Based on these awards, Mr. Powers said, "We found that in Year 8, boys and girls are pretty much equal award winners. Year 9, Year 10, Year 11, and Year 12...not so. More girls. Almost an embarrassment." Though awards are sometimes a poor indicator of real achievement, they were concerned about boys not seeing themselves as high achievers, and this concern was brought to the Boys Working Committee. There was a sense that these indicators were contradictory: OP results indicate boys are fine, but they get fewer awards. How can that be? These are not necessarily contradictory, though, for these too, along with the disciplinary disproportion, stoked a growing belief among certain members of the working committee that boys were not "naturally" behind; instead, boys were being treated unfairly.

As is typical of Springtown, the members of the working group decided to democratize the process and seek out the concerns of the school's other teachers. They held a meeting after school, the results of which were disheartening for the two leaders, Mrs. Sanderson and Mr. Tully. Sanderson said of the incident:

> We tried to get it [the boys' education initiative] off the ground last year, but it was sabotaged by teachers who only wanted to talk about boys with bad behavioral problems as opposed to "How do we proactively help [*inaudible*]?" And then the other issues that were going on, you know fathers not being in households and that sort of stuff, but it was just ransacked by teachers who— No men [on staff] came to the meeting, let me tell you, no men at all...It was all women who were having trouble with coping with the boys in their class. So they were all there to bitch, basically. And so we saw sort of quickly it just wasn't going to work.

The story that both teachers—both committed to and motivated about the issue—tell about this meeting demonstrates the tensions that can occur over boys' education (and other politically fraught issues) when they arise in schools. Rowan and colleagues refer to the "dangerous territories" of reform targeted at boys, and Kenway and Willis show concretely how tensions can develop amongst staffs that regard such issues differently.[11] These two teachers ran afoul of another group of teachers who were not as committed, or who were actually resistant, and the experience was "disenchanting" and left them feeling "sabotaged" and "ransacked."

Additionally distressing to both Tully and Sanderson was the lack of progress of the Boys Working Committee. Their solution to this sense of rejection and stalled progress was to forge ahead on their own with plans to work with the "hard nut" boys. They organized the father–son fishing trip, created building projects for commons spaces, and took time "catching them [boys] out in their own environment" to talk with and counsel them. Initially, though, this work was simply added to an already heavy responsibility load for these two, both of whom taught regular classes and had leadership duties. Eventually the administration granted them more release time for this work, but the strain of their workload was clear for both. They seemed to crave other teachers' support, both as affirmation of their concerns and for help so that they wouldn't have to squeeze in "one more thing" to their duties.

Despite the gender-specific concern the Boys Working Committee began with, the explicit focus on boys was not to last. Shortly after the meeting with teachers described above, the administration decided to change the group's name, from the Boys Working Committee to the Millennial Kids Group, reflecting also a new focus on generational issues—based on the work of Howe and Strauss[12]—rather than gender issues. Howe and Strauss's formulation posits cyclical generational attributes and heaps praise on the potential of the generation in high school around the turn of the twenty-first century. The various stakeholders within the Boys Committee had differing explanations for changing the focus to Millennials. Mr. Powers, for example, said that the group members were trying to avoid the swings back and forth from girls' education to boys' education.

> Now in some cases it will be boys that we focus on. In other cases it will be much more neutral in terms of gender, and maybe more specific to the generation. And that seemed to have a much more— It seemed to gel more with a coed school and our philosophy here of inclusiveness than to go in there and you know sort things out simply because there's a particular group that needed help at the expense of others. Now, that's why it changed. Now, not everybody's happy with that because they do see a need to fix up the issues surrounding boys.

His explanation sets the founding principle of coeducation as reason for the change. He also suggests that the move was an effort-saving tactic, avoiding the endless cycle of taking care of one gender and then the other. Mr. Tully was less sure of the reason for the change:

> [*Sigh*] It was sort of— There is a lot of mixed stuff going on there. I mean, Mrs. Sanderson and I did a bunch of research for that, but [*pause; sigh*] I am just not sure what was going on with agendas in that because we did a lot of research and I thought we were actually going to get some stuff moving and get it moving fairly quickly, and then I felt it sort of fell down. It sort of lost a little— I think it lost a lot of its power.

Whatever his uncertainty about the group's ineffectiveness, he suggests that "agendas," in a pejorative sense, were the cause. Mrs. Sanderson was less subtle in her critique:

> "Millennial Kids." OK, that's an interesting term. [*laughs*]. It was the Boys Committee, because we were very very very concerned about our SMT boys.... And then Mrs. Howard approached me and sort of like said "You know, I'm just a bit worried if we do something which talks about boys in education. We should do something that's about boys *and* girls in education. But in the middle stuff, maybe you can focus on some strategies for boys." So it was like watered down completely. So we just lost interest in it.

Mrs. Sanderson later talked again about the "Millennials" focus, suggesting both that the decision to change focus was not entirely democratic and that it was a perversion and weakening of the needed focus on boys.

> I still like to think [Mr. Tully being put in charge of boys' issues] is, to me, the first thing that's been done towards assisting boys. Because it's not about "millennial kids," and let me tell you how that word came about, because Mr. Tully and I were so against it. [*laugh*]. Mrs. Howard read this article... and it was all about millennial kids and how they're computer age. And it was just decided that our policy would be about millennial kids. It was no longer a boys in education policy. So, it's about boys, but we're not allowed to use the word "boys" in case we offend girls and women.... We've got to be able to call a spade a spade. Because if you don't, you lose sight of who you're actually trying to identify.

Sanderson suggests that there was much fear, justified or not, perceived in the reaction to the boys' education focus—fear for the reaction of the school community and for the culture of the school. What resulted for Mrs. Sanderson was once again a feeling of being "sabotaged."

When, two years later, I posed to administrators that there had been a feeling among some members of the group that the change had been "forced," they clearly took exception to the notion. As Mr. Powers asserted,

> Now, I have to disagree with the word "forced" because the committee of which Mrs. Howard and I were on, I suppose there were two perspectives on it, and I think Mrs. Sanderson's perspective is what you've got there and not probably the group's perspective. She was very definite and vehement that it had to be about boys, and I think in the group she may have been the only *one* who felt like that.... Yes, I'd say that was more—*Mrs. Sanderson* wasn't keen on the concept [*laughingly*] as opposed to the group, and Mrs. Sanderson's not here anymore.

Indeed, Mrs. Sanderson left Springtown. There is no way to know if this incident was a part of that decision to leave, and I am not suggesting it was, but her dissatisfaction was apparent when I first interviewed her. She felt like an "outsider" and a "troublemaker." The raising of boys' education issues, as "dangerous territories," tends to create such marginalized positions for people.[13] Springtown was no exception.

I don't recount this story to claim what is the "Truth" about what happened or to lay blame and castigate particular people. As Mr. Powers rightly points out, there are differing perspectives on both sides, and one is no more "right" than another. My point is, rather, to show how such tense issues, which access core beliefs about gender, fairness, professional autonomy, and child welfare, can be approached and how, as a result, they can fracture a staff. Even at a progressive, open, democratic school like Springtown, such issues present dangers and strong emotions.

Ultimately, too, the difficulties of creating a boys' education program at Springtown came down to a clash with the school culture. The foundational principle of coeducation was endangered by focusing on boys, for Mr. Spears was always quick to point out that their initiatives that began as boys-focused were always open to girls too. Despite the nominal change from boys to more inclusive programming, which Mrs. Sanderson believed "watered down" the focus on boys, many key initiatives continued for boys to target their needs. The nominal change was crucial, though. The ethos of the school could not allow otherwise.

What We Can Learn from this Case

The shift from the boys' education working group to the millennial kids group provides important insight into the boys' education efforts of SRS. Even for a school with policy independence from both its state and denomination—meaning it could do pretty well as it liked in regard to initiatives for boys—a complex picture of reform developed.

A primary influence on boys' education reforms at Springtown, as I have stressed throughout, is the creation and preservation of school culture. Again, this made the school administration wary of new programs that challenge core concepts like coeducation. This directly changed the focus of the boys' issues working group. Still, because Springtown attempted to interrupt status quo gender relations in certain ways while leaving many traditions and gender roles alone, what they've developed is a complicated mix of the progressive and traditional. They're coeducational in a highly single-sex dominated sector, and they practice little gender segregation in any explicit way; there does, however, seem to be a high degree of gender segregation in peer groups and informal settings (during tea, etc.). Outdoor education instead of competitive sports lessens the impact of masculinist hierarchies and allows boys and girls to see different sides of one another, though some interactions during outdoor education demonstrate sexist treatment of girls (boy–girl groupings; boys too quick to do things for girls). Some teachers, according to my focus groups with students, actively discourage girls from taking part in traditionally "male" subjects and actively discourage boys from taking up "female" subjects. The uniforms and appearance policy reflect a strong sense of gender-appropriateness, for boys are not allowed to wear long hair or earrings, a matter about which several young men complained. The administration is cognizant about the need for gender balance and often ensures equal representation in public events, and female teachers and administrators take visible leadership roles in the school. The ethics of caring and cooperation at the school provide boys a different model of relating that's in stark contrast to the competitive, hierarchal models of other private schools. Also, there appear to be a great many non-romantic relationships between boys and girls at the school. Yet Springtown seems, in many ways (though with exceptions), to follow a pattern of gender-based divisions of labor, with subjects staffed in traditional ways and the emotional labor somewhat weightier for female staff. These facets of the school culture, again, suggest a complex mixture of progressive and conservative gender beliefs. This also sets up a complex internal political dynamic in which those inclined one way or another about boys' issues are pitted against one another, which has in some ways fractured groups of teachers and alienated those who felt themselves to be on the losing side.

Another key finding was that Springtown used the *BGIR* report extensively, though not in any straightforward way. Mr. Spears had a copy with him during our first meeting, and he had shared the results with the Parents and Friends Council. There was a link to *BGIR* on the school's Web site (on the parents' resources page), even until 2007.[14] Mrs. Howard used the report and some submissions to it in her report on resilience and in formulating new disciplinary interventions for boys.[15] Nonetheless, the use of the report by individual teachers was sparse. As the administrators explained,

> HOWARD:... if you were to go out there and say to ten teachers "How you see this report and the recommendations being translated into what happens in this school?" they'd probably say, "Well, I'm not really sure." But if you said, "What are your attitudes, values, and beliefs about boys' education?" they would probably have a whole lot to say,... So it's [*BGIR*] filtering through in not as formal ways, but it's probably—

SPEARS: But teachers don't read reports such as this, in that kind of format. You know, that [professional development and conference updates] is how it comes back into the school.

Part of why *BGIR* comes in through these alternative channels is based on administrator perceptions of teachers' needs.[16] The danger, of course, in relying on digests of research and research recontextualized for consumption by teachers is that these often strip the political context away from the original research.[17] Such digesting practices, though, are sometimes a necessary reality in an era of profuse, globalized discourses on topics like boys' education. Mrs. Howard, in explaining Springtown's approach to *BGIR*, encapsulates the difficulties faced: "the report is one part of this whole big jigsaw. And I'm just trying to pull a few themes, you know, out of it that will have meaning for us." *BGIR* really is only one part of a large jigsaw within an increasingly complex picture of adolescence, development, school culture, social and technological change, accountability frameworks, and so on that administrators must negotiate. Any examination of such a policy-report must keep that in mind.

Springtown is also a good example of what happens with issues of boys' education within economic competition between schools. In Mr. Powers' interviews, he gave much attention to how Springtown compares with other schools. There can be little wonder that their concerns with boys revolve in great part on the *noticeable* indicators—those the public pays attention to—like behavior, awards, and OP results. Not only is this a marketing concern, but it reflects exactly the challenges of triage and competition in a neoliberal context where, to survive, schools must attract students and, more especially, their parents.[18]

Finally, it would be remiss not to note the key role the administration plays in the progress or limitation of boys' education at SRS. Though my research admittedly focused much more on the leadership than on, say, classroom interaction, clearly the administration acts in the capacity of "key mediators" or gatekeepers of policy.[19] They take a structured role in interpreting policy for others, and they are the protectors of the school vision and ethos. The possibility of cultural atrophy is always present when held in the hands of a few key players, particularly when those players leave, but for Springtown it seemed that such vigilance actually preserved the progressive parts of the culture, not really the conservative parts. Though there are potentially quasi-coercive measures available for enforcing their interpretations and their versions of the school culture (like, say, the teacher evaluation process), the leadership team rarely used such means. To the contrary, much of the power of the administrators is derived from a truly open spirit of collaboration and professionalism. As teachers and students alike told me throughout the research, members of the community honestly felt they had the license to speak their minds. They may not get their way, but they felt they would at least be heard.

What the Two Cases Tell Us about Local Approaches to Boys' Education

One can surely not extrapolate from two small cases, regionally located, to make grand pronouncements about what boys' education reforms will look like in all schools. These cases, paired with one another, though, shed light on *some* ways that *some* schools might approach this divisive, politically tense issue. Particularly because these are local cases, and because in most contexts (like the United States, as I explain in chapter 6) boys' education reforms are highly localized, these cases are perhaps the best window we have for assessing the transfer potential of boys' education reforms. In this final section, I point out what these two cases together suggest about boys' education. How and why do people develop interest in boys' education? How do they persist? How do they resist? What role do federal policy and initiatives really play in the interventions that happen at the school level?

The Uses of *BGIR* and Other Materials

First, I want to explore a key "how" of the approaches of these two case contexts: how did they use resources, such as the focus of this book *BGIR,* and other information? This question is crucial to understanding the political dynamics of how these educators went from a field of contested knowledge—the debates over boys' education—to daily practice.

No one I talked to at any Australian school, conference, or educational gathering ever suggested that they had no prior knowledge of boys' education issues. At the very least, all had seen media stories about boys; such stories were nearly inescapable in newspapers and on television and radio. A coterie of popular-rhetorical boys' consultants and authors (such as Lillico, Fletcher, and Biddulph) were constantly touring the country giving workshops on boys' issues. Their popular press books were the most accessible sources, ones that could be easily found in local bookshops, ones that were advertised heavily, and ones that were written in language geared to parents, teachers, and administrators. These were also the sources most heavily reflective of traditional, conservative gender beliefs. More nuanced, academic texts were hard to find, expensive to purchase, harder to read, and sometimes challenging to certain people's deeply held beliefs about gender roles. The government, too, was of little help. Before *BGIR*, which appeared less than a year before the RSC began its project, the Commonwealth had not created any policy on boys, and, as the epigraph that begins this chapter suggests, the states and territories had given little direction. This left schools in a position of utilizing popular-rhetorical works as de facto policy.[20] In Springtown's case, *BGIR* eventually became, upon its release, one of many texts from which they drew for policy.

Most teachers and administrators in these case studies did, in fact, read texts that discuss boys' education. Several participants from both case studies showed me large binders filled with articles about boys' education they had collected and marked up. Many talked at length and in detail about books they read, classes they took, and conferences and seminars they attended, and each was able to provide reflective critiques of these sources. Mrs. Hirst, from Gretna Green, for example, explained

> I've done a lot of reading; any conferences I've been to I've followed up with professional reading. I was very interested to read one of Martin Mills' students—was here doing a prac[ticum] and she gave me a book of readings that she has for the subject, and I've read some work of his. I've read those—I've read all the rubbishy ones from Lillico and Biddulph and I see those books as being really dangerous in the way they perpetuate these ridiculous stereotypes that to be a real man you have to fish and crab and hunt. I mean it's quite ridiculous.

Clearly, teachers and administrators don't often come to boys' education issues in an unreflective manner; it does happen, but it was not present within the participants of my case studies.

My biggest concern in this book, of course, is the use of *Boys: Getting It Right*. The report was largely unused in the RSC and unevenly used at Springtown. In my interviews with the teachers and administrator in the RSC's BELS project, nine of the fifteen participants had some knowledge of the report's existence, but *not one* had actually read it. Thus, they didn't use the report itself, though many could have known the report's basic findings from the media without knowing they came from the report. At Springtown, several people used the report more extensively, and most who had read it could recall details and talk at length about its relative usefulness.

Perhaps the prototypical way that the participants of my study utilized *BGIR* and the texts available to them was a process of "cherry picking"—selecting only the elements that fit their existing ideas, that they found appealing, or that seemed innovative or important. Mrs. Howard's metaphor of a jigsaw puzzle, quoted above, is an apt metaphor, for educators often pull from many sources trying to put together a complete picture of what boys are like and how they should be educated. Since *BGIR* is largely symbolic and deregulatory, simply steering at a distance through its resulting initiatives, there is little reason for schools to follow every edict. Instead, they're left to "pull a few themes" from the report that fit their existing needs.

Also a key issue, teachers largely don't use the boys' policy resources in their original forms. As Ball notes and my research confirms, many policy texts are never read, even when they are foundational curriculum documents.[21] Rather, key mediators are relied upon to translate policy for those educators. At Springtown, administrators performed this mediating and translating role, but for many others the popular (and, one must say, conservative) media was often left to do this.

Why Boys' Education?

Another key question to ask of these case studies is, Why do people initially address boys' education? Is this some inescapable discourse that entangles them, or do people arrive in more complex ways? Are they forced by circumstance, enticed by real interest, driven by zealotry, or pushed and pulled by political ideologies?[22]

Based on these case studies, and my research generally, there are several key reasons why educators and schools take up boys' education. First, many who try such reforms have experience or genuine beliefs that boys need help. The RSC's Mr. Barr puts it most baldly: "My interest in this is based from the boys, knowing there is a problem." Whether accurate or not, educators and service providers come to boys' education because they perceive that the boys in their lives are not doing well and they want to fix it.

Second, for some, curiosity brings them to boys' education reform. They might be curious about what everyone is so concerned about, or they might be curious to find more effective ways to reach boys than the simplistic suggestions given by popular-rhetorical books. Mr. James, from the RSC, for example, says

> I didn't think I knew as much about it as I could. Besides those few seminars that we went to last year—I think I went to two after school—this was basically the reason I got into it, because I had only those tips for teachers type of hints: dim lights, bright colors, et cetera, et cetera, et cetera, short, sharp activities.

In general this curiosity is driven by a practical desire to be able to do more for the students in front of them. It might also help them deal with parents better:

> ...the parents have come and said "This is what we've done [with our sons]. What else can you suggest that we do?" And I find it a really hard thing to answer, because I'm not really sure that *I* know what to do in my class. (Mrs. Elliott, RSC teacher)

Thus, for some, getting professional development on boys' education issues solves the real needs they have for daily survival for themselves and their students.

For others, though, particularly female teachers who might regard "being a boy" as a nebulous but pedagogically powerful experience that they want access to and understanding of, professional development on boys' education might seem to offer answers. Consider what Mrs. Cargill, an RSC teacher, had to say:

> I think probably just that I wanted to be able to relate to them [boys] better and especially boys that were having problems. Perhaps I thought I might be able to find out some strategies to just make get them feeling easier and more successful and get them to be more interested and try to think a bit more on their level, not having had grown up boys or even boys that are their age, just learn a bit more about how they ticked I suppose.

Though problematic to assume that boys "tick" in ways necessarily different from girls, many remain concerned that teachers need to understand boys better if boys' outcomes are to improve. In some ways this concern fuels a backlash perspective that posits that boys are "thorns among roses," disadvantaged by female teachers who don't understand their "true" nature and who punish them for not being like girls.[23] Still, the impulse to know a group different from oneself is easily understood.

Third, boys' education reform, for some, could be used instrumentally to improve other aspects of their schools or students besides just boys. Some teachers believed, for instance, that learning techniques put forth to help boys would actually help all students, girls included." For such teachers, professional development was often about finding techniques that could *broadly* improve their situations, not necessarily just on the explicit focus of the professional development.

A fourth driving force toward boys' education is money. Though it would perhaps be too cynical to assert that anyone would attempt gender reform *solely* for grant money, clearly the funds' availability had a determining effect on reforms. Springtown invested a good deal of funds in boys' education even though their reform was limited. They had money of their own to fund attendance at expensive conferences, to put on fishing tournaments and building projects, and to give Mr. Tully release time to work with boys. In a context like the RSC, which had only the BELS funds, the options were more limited. Most importantly for them, when the grant money ran out and they failed to get the phase two funding, the formal boys' education reforms largely ended. The interest didn't disappear, but the cluster moved on to things they could afford. In a context of chronic underfunding of public schooling—even while most private schools flourish with "extra" government payments—interest and initiatives follow the funds.

Fifth, another aspect that brings some people to boys' education issues (and many other "hot topics") is that the professional development elements involved offer educators opportunities for rest and professional networking. More than one teacher conversing between sessions at the B2FM conference told me that they didn't care much about boys' education issues; they were there "for the day off" and the fancy snacks and alcohol. The RSC's BELS participants also noted that, to them, the project's most gratifying aspect was getting to know the other teachers and hearing their similar experiences and worries. Though it may, to some, seem superficial to consider these bodily and social pleasures, it is important to realize that such motivations are strong and consequential, a reflection of the reality of teachers' labor being physically and emotionally difficult, often undervalued, isolating, increasingly de-professionalized, and constantly intensified.

A sixth driving force, one that might similarly seem self-concerned to some (though why should it?), is that boys' education presents opportunities for professional advancement. This is clearly true for the many consultants that flock around boys' education issues. This could also come true for individual teachers. Being an expert in a topic makes one more valuable within a school, which might garner a title, more pay, or possibly extra time out of the classroom. For Mr. James, a new probationary teacher in the RSC, "I also did it because I thought it might have given me a little more longevity here at the school." For him, signing up for boys' education reform was an attempt to establish job security, a way to show his motivation, engagement, and willingness to help his principal.

The final motivation for engaging with boys' education issues I discuss here (others are possible) involves external pressures on educators. Policy was not really a coercive force in this case, for much of *BGIR* and its initiatives were voluntary and largely symbolic. This was particularly true for Springtown, which has policy independence. For the RSC, though, the grant funding acted as an externally motivating force. It imposed deadlines, ones that, with the change of principals, undermined the project's intentions. There was also a sense that the money was a constraint because it *had* to be spent by a certain time or it would disappear. For individual teachers and principals, there were similarly external pressures. For some, the pressure, particularly when coming from those in power, was unpleasant. Two teachers from Sunshine Valley (RSC), for example, explained how they got involved in the BELS project:

> MARCH: Want us to be honest?... Because we were told we had to.
> McLENDON: We had to.... And probably because we went to that thing [a boys' education seminar] last year. But again that [the seminar] was pretty much a case of [an administrator saying] "You're going."

While few put it so plainly, several RSC teachers noted that their participation was not completely voluntary. In all, half the teachers were specifically asked by their principals to participate, regardless of any personal interest in boys' education. This was not always the case, though. Some were actually eager to participate. As Mrs. Dalloway told me, "I actually tried to volunteer three times, but because I'm [a] Year 3 [teacher] the principle kept knocking me back. So eventually no one else would volunteer, so they had to take me."

The main point I am making is that people come to boys' education for many different reasons, probably the least of which is passion or zealotry—though some, like Mrs. Sanderson, did. Most educators who engage the issues surrounding boys in schools come with honest concerns and motivations, most altruistic but some self-concerned. Progressives must attend to such reasons, for they hold insight into how they might enlist educators and service providers in positive, progressive, and equitable gender reforms in schools. I take up that topic more explicitly in the Coda.

Why Not Boys' Education?

A third key consideration, beyond the use of *BGIR* and the reasons people take up boys' education, involves asking Why do educators abandon or *not* take up boys' education? Knowing this can help in developing programs that have longevity and lasting impact on students, teachers, schools, and communities.

One common and perhaps obvious reason for abandoning or never taking up boys' education reforms comes from experiences of failure with programs. For some, trying to avoid the roller coaster of reform from one topic to another—a distaste for "fads"—was enough to make them want to avoid boys' education issues. The metaphor of a pendulum, constantly swinging from panic over girls to panic over boys and back again, is common in teacher discourses, and it has a limiting effect for many educators. It leaves them, though, with a notion that the status quo or, more aptly, a notion of "gender neutrality" will keep them from having to go back and forth between boys and girls.[24] This belief, though, is a dangerous position for women and for some boys (based on socioeconomic status, race, sexuality, and so on); it just makes their problems invisible or unmentionable. Nevertheless, it is easy to feel sympathy for teachers suffering reform fatigue, who have seen topics come and go more times than they can count; the danger lies in rejecting issues out of hand when they may be more than just swings of the pendulum.

Other teachers have actually tried programs in boys' education and have found them lacking. I heard several teachers talking of their schools trying something for boys ten years ago that didn't work, so they wondered "Why try again?" Teachers at Springtown, a school culture founded on reflection and professional experimentation, were able to move past such setbacks, but teachers in other contexts may not have the resources to avoid such pitfalls.

Allegiance to girls programs and concerns was a second reason for skepticism or resistance to boys' education issues. Mrs. Dalloway, a teacher in the RSC, for instance, recounted that

> I've been teaching 21 years, and I went through the feminist movement in teaching, how we were actually favoring the boys and I actually analyzed myself years ago—how many questions I asked the boys and how many questions I asked the girls—and discovered I was having a problem back then.... I didn't really, really think that there was a problem with boys at all except that they misbehave and, you know, there are ones like in my class at the moment who I can't really get a handle on.

Ultimately, Mrs. Dalloway's experiences with girls' education reform didn't prevent her from pursuing boys' reforms, but other teachers I met clearly had their minds set against taking any focus from girls' needs.

Third, for some, avoiding social divisiveness was reason enough to avoid boys' education reforms. Mr. Barr, a primary teacher in the RSC, used the metaphor of a "poisoned chalice" to describe this impulse. "[B]eing seen to be doing too much for the boys" makes for a dangerous, divisive situation. This seemed to underpin much of the reluctance to do boys' education work at Springtown. Ironically, though, the attempt by administrators not to be seen as unfair to girls itself caused rancor and division within the faculty, at least for a few.

Next, just as the presence of money brings people to consider boys' education issues, as mentioned above, lacking money or having funds run out can lead to the obverse effect of abandoning or not pursuing boys' education. Springtown, as noted earlier, had more resources to expend. The RSC only had funding insofar as the BELS grant allowed; there were no "discretionary" funds. Then, when the funds dried up, so did the work on boys' education.

Another tight resource in schools is teacher time and energy. Some teachers, particularly those with heavy responsibilities or the "star teachers" who take on every program, may avoid or drop boys' education initiatives because these are just "one more thing to do." For Mr. Tully and Mrs. Sanderson at Springtown, this seemed to be a salient factor. With the resistance to their initial forays, both seemed to retreat from the whole-school initiatives they originally envisioned, turning instead to small things over which they had total control.

As I noted extensively in the case of Springtown, conflict with school culture and traditions can also be an important point of resistance to boys' education. Springtown was not alone in this, though. For other schools that have a strong ethos built on coeducation or other principles, there are concerns that boys' education could conflict with how they want to approach schooling their students, so they avoid it.

Finally, just as there were external pressures that pushed some people *toward* boys' education concerns, some external pressures push people *away* from those issues. Certain policies might limit what schools can do in boys' education, particularly if they felt the existing *Gender Equity Framework* or the Sex Discrimination Act were being violated. At the chalkface, though, the external pressures were again more from immediate supervisors or from funding concerns. Mrs. Sanderson and Mr. Tully at Springtown felt that they were receiving pressure from administrators and even other teachers to stop focusing on boys.

In summary, much like the diversity of reasons educators and service providers come to boys' education reform, there are many diverse reasons some educators drop or never pick up boys' issues. Many of these are complementary, suggesting that the presence or absence of certain things may have a determining role in whether or not educators engage in boyswork. Thus, if there is, say, money for boys' issues, educators are *more likely* (not "certain") to take them up; without money, they're more likely to drop such issues. Whatever the case, such understandings can be useful in developing programs that last and that move in progressive directions.

Implications for the United States and Elsewhere

From my experience as a high school teacher, a teacher educator, and a researcher, educators in the United States are not much different from educators in Australia. They have similar challenges, similar motivations, and similar tasks that they perform daily. The big difference, of course, is the political context within which they work. There is no specific U.S. boys' education policy to implement, though *No Child Left Behind* (*NCLB*) legislation is an inseparable part of the lifeworld of teaching there. *NCLB* is very different to react to, though, because it is highly regulatory, not symbolic like *BGIR*. Teachers are less free to cherry-pick the parts they want to take on.

Teachers in the United States, because there has been no federal or state policy to guide them, as I detail further in the next chapter, have had to create local, diffuse solutions to solve their difficulties with boys. They have looked to authors like Gurian, Pollack, Sax, and Sommers for answers.[25] They have looked to seminars and workshops, like the ones I describe in the last chapter and the Coda. They chase grants. They also worry about the competition from private schools and other school choices that have proliferated in a time of increased accountability, testing, and neoliberal managerialism. Largely, though, individual teachers and schools solve their problems on their own.

The results of these searches for workable solutions are often as regressive, conservative, and stereotypical as the dominant discourses that circulate on boys' educational needs. The *Washington Post* featured attempts to create reading "tribes" for boys, putting them in boys-only classes where they do hands-on activities and dance to the Village People song "YMCA," because their brains are supposedly different from girls' brains.[26] *Teacher Magazine*, a mainstream national publication, detailed one teacher's program to fix the "blurred gender lines" of the modern world.[27] As the author describes the knights and dragons activities:

> Using their fine motor skills, the girls worked on creating headdresses and decorating their castle. The boys used large motor skills to create huge dragon posters that they helped to string up between trees.
> The culmination of our efforts was an outdoor dramatization that students and teachers in other classrooms envied.... They were interested to see the knights use all their energy to slay those terrible dragons with beanbags. And when the knights fell to the ground, a group of beautiful young ladies rushed to their sides with healing herbs from the pouches they had created.

Quite frankly, I find such blatantly sexist practices disheartening. Clearly, some have taken the spaces created by the lack of policy in the United States as an opportunity to install discourses of—and practices based on—biological and brain differences, fused with conservative and religious discourses of "appropriate" or "Godly" separation of the sexes in schools. These "commonsense" and quasi-scientific notions are rife in the

United States, but I would venture that other countries, too, have seen this. These are the most *available* discourses worldwide, but they are deeply problematic.[28]

Monitoring such situations on the ground in some contexts, as I have previously noted, can be very difficult. As I show in the next chapter, the U.S. situation, for one, is highly localized, and because of the population size and widely different contexts from state to state, it is often difficult to know what is being done under the banner of boys' education. Nevertheless, it is imperative that such pedagogical work be examined and that progressive, well-researched, and practical discourses be made available to educators. I return to this point in my Coda. First, though, I want to focus exclusively on the United States, both as an example of another country wrestling with policy on boys' education itself and as a comparative case that underscores important lessons of Australia's policy on boys.

Chapter Six
Boys' Education in the United States: What Australia's Example Tells Us

When I went to cocktail parties during my research year in Australia and told people I study the education of boys, normally I got stories about my interlocutors' own sons or a newspaper account they read recently about boys. At the least, there would be a look of familiarity, or sometimes suspicion until they figured out on which "side" of the issue I stood. In contrast, when I went to cocktail parties back in the United States and told people I study the education of boys, there was normally a moment of silence, the cock of one eyebrow, and the inevitable question: "Boys?! What's the issue there?" Things have changed in the United States—boys' issues are *much* more prevalent and people are more cognizant of them—but there is still a qualitative difference between the two countries. Why should this be?

As I described in chapter 1, in many countries—not just Australia—similar concerns have been expressed over boys and their education. Then, throughout, I have described how Australia's world's first federal-level policymaking on boys' education resulted in a lengthy report, *Boys: Getting It Right* (*BGIR*), and several national initiatives to create a research base and praxis. Australia isn't alone or unique in this level of national attention, though. In England, for example, articles on boys appeared frequently in the *Times Educational Supplement*, and books targeted at teachers for raising boys' academic achievement have flooded the market.[1] Also, numerous local educational authorities have commissioned reports to consider boys' issues.[2] In Canada, similar reports have been commissioned on boys, and in New Zealand the right-wing National Party has made repeated calls for a Federal inquiry into boys' education, and a media panic over boys in 2006 brought much comment from the minister of education.[3] In Iceland, boys' education discourses similar to those in England are prevalent among teachers, and some universities are teaching classes especially for training male teachers.[4] These are part of what I earlier noted as a general "boy turn."[5]

What about boys' education issues in the United States, though? Despite the high visibility of debates internationally, issues of boys and masculinity are still relatively nascent discussions in the United States. High-profile exceptions and certain movements around boys' education have been present for years, but the United States has yet to experience the concerted governmental and public panics and policy that has characterized the topic elsewhere, particularly not to the extent as in Australia. No U.S. Federal Government reports. No inquiries. Rather, stated simply, *the context for boys' education issues in the United States is diffuse, conservative, structurally and legally constrained, and localized.* Finding causes for this situation requires exploration of the complex circumstances—the ecology—of U.S. education.

In this chapter, then, I answer this overarching question: Why has the United States not seen panics and policy movements to the extent Australia and other countries have? Certainly many have expressed concerns about boys, but why have such movements been less "successful," less part of the consciousness of the general public? (Teachers are a different matter.) What differences between the context I described in Australia and that of the United States might explain this?

In this chapter, I return to a policy ecology metaphor to explore whether and how boys' issues have been or might be addressed in the complicated, disparate, structurally determined U.S. context.[6] Just as I did with the Australian example, I start by outlining the history of gender and education in the United States, connecting it to emerging concerns over boys in the popular press . From there, I explore specific issues regarding U.S. boys that are often used to argue their needs. Despite these emerging anxieties, I argue that numerous factors either prevent large-scale interventions and policy, or these factors quarantine interventions at the local level. Then, I examine some successful (in a progressive sense) and high-profile U.S. cases of educational work for boys, particularly African American boys. Finally, I consider current momentums within education and government that could cause boy debates to gain wider attention in the United States in the future, and I sound a few cautions about these possibilities.

Gender and Education: A U.S. History

The United States has long had public debates over education and gender. Most controversial throughout this history has been coeducation. In the last third of the nineteenth century, as girls began entering public schooling and consternation over the ills of coeducation followed, educationalists began seriously considering gender. Much early work was critical of coeducation for students' physical wellbeing. Edward Clarke's *Sex in Education* is perhaps the most famous example. Clarke decries the threat to girls' "catamenial functions," or reproductive abilities, from excessive usage of the intellect during coeducation. Not so for boys; in fact, he considered exercising the intellect to be a good way to redirect boys' biological "urges."[7]

Anxieties were not only expressed over the presence of female *students*. As Gail Bederman recounts, boys' contact with female *teachers* was also a cause for worry, made particularly clear in the work of G. Stanley Hall during the turn of the twentieth century.[8] Hall recommended that elementary teachers let boys act like "savages" to counteract the physical weakening from female teachers' undue civilizing influence. Similar panics over the "feminizing" of boys by women teachers have never totally gone away. From Patricia Sexton's influential polemics on the "decline of manliness" to recent diatribes against "progressive" education's attempts to make boys more like girls, arguments blaming female teachers for boys' difficulties continue apace.[9] Responses to the perceived crisis—efforts to re-masculinize boys—have taken differing forms, including the use of physical education, boys' organizations such as the Boy Scouts, and even military training.[10] Contemporary boys' programs cannot help but confront or be seen in light of the continuing legacy of both fears over demasculinization by female teachers and efforts to instill a manufactured, traditional masculinity in boys. Put simply, moral panic still characterizes gender in education issues.

Still exhibiting this crisis character, a turn to girls' educational issues began in the United States in the early 1970s (about the same time as in Australia; see chapter 2). The successful 1972 passage of Title IX, the law prohibiting discrimination by sex in institutions receiving federal funds, both capitalized on and created renewed interest in girls' access to schools and educational resources, most visibly in athletics. At the same time, closer scrutiny fell on the representation of females in the curriculum and the treatment of girls in schools.[11] The Women's Educational Equity Act soon followed, and a resource center for improving girls' educational experiences and achievement was created. Attention to girls' issues ebbed and flowed in subsequent years, but the early 1990s saw a firestorm erupt over girls' issues again.

The catalyst for the renewed crisis was the American Association of University Women's (AAUW) report, *How Schools Shortchange Girls*. In it the AAUW charged that U.S. policymakers left out girls' issues in the educational reform flurry following the *A Nation at Risk* report.[12] The neglected issues, they argued, include girls' declining self-esteem, gender biased testing, math and science achievement gaps, and the absence of women and women's issues in the curriculum. The U.S. media quickly promoted the report, often characterizing it as a first glimpse at a major crisis. Other research, both from the AAUW and others, built on the report's themes, emphasizing girls' depression and low self-esteem and the role schools play in these.[13] This again resulted in legislation; the U.S. Congress in 1994 passed the Gender Equity in Education Act, largely in response to the renewed attention to girls' issues.

Starting in the late 1990s, a distinct shift in attention toward boys' educational issues began in earnest, but no significant legislation has yet been proposed or passed. Largely, concentration on boys has come from a spate of popular trade books addressing the topic of raising boys and polemical texts about alleged failures and deceptions of feminist educational reforms. These include Pollack's *Real Boys*, Gurian's *A Fine Young Man*, Kindlon and Thompson's *Raising Cain*, Thompson's *Speaking of Boys*, James Dobson's *Bringing up Boys*, and a slew of others. These tomes on raising boys have been the most visible expression of concern; several, in fact, have made the bestseller lists. They're also largely conservative in their political orientations.[14]

A particularly important example of conservatism is Christina Hoff Sommers' *The War Against Boys*. Sommers, a long time critic of left-leaning feminism, charges in the book that "misguided feminism," often through calculated deception, has turned the educational system counter to the "natural" proclivities and needs of boys.[15] She proposes fixing this by "returning" to moral education and "basic values, well-proven social practices, and plain common sense" (209); this represents a neoconservative wished-for return to traditional masculinity.

Books that focused on teaching practices for boys have had similarly conservative, stereotypical arguments.[16] Michael Gurian, for example, like Sommers, critiques progressive education in *Boys and Girls Learn Differently!*. In his view, educators have forgotten or ignored the "natural," brain-based differences between boys and girls, to the detriment of both. While some of his recommendations address learning-style differences that could indeed be considered brain-based (pacing of lessons or using manipulatives, for instance), many cross the line into highly social interventions like "rites of passage" experiences, character education programs, and school uniforms (not typical in U.S. public education, though increasing).

Several points should be considered about these books—and others like them—as central players in the U.S. context for boys' education. First, they're often practice-oriented, and relatively few theoretically oriented works on boys' schooling and masculinity have originated in the United

States.[17] This is key, for U.S. boys' education debates have thus primarily taken place at the popular rather than academic level. Popular books on boys encourage parents and educators toward activism, but they elide the possibility of structural and political change. Further, the texts' practical foci and ease of reading—and the absence of response from education research—have effectively distributed *one particular version* of boys' problems. Worse, in an era of site-based management absent of state guidance on boys' education, such books have become "de facto policy" for schools, just as happened in Australia.[18]

Second, the books' conservative nature deserves scrutiny. Most rely on stereotyped or biological views of masculinity. Many, like Sommers, are explicitly antifeminist. Most also appeal to "common sense" and a neoconservative desire for "simpler times" when people "knew their roles" in society. This is all dangerous material on which to base policy.

Still, the issues faced by boys in the United States, just as elsewhere, are not *only* manufactured crisis, backlash against feminism, or a bid to return to religious orthodoxy. Parents and educators are not dupes, after all. Many negative indicators of boys' achievement, social status, and health have been cause for major concern among well-meaning people.

Where the Boys Are: Issues for and about Boys in the United States

Most issues that are of concern for boys in Australia and elsewhere have also been of concern for boys in the United States. Academic issues in U.S. schools have, as they have internationally, focused mainly on literacy, though boys also lag in other subjects. In literacy, boys' scores trail girls' in the National Assessment of Educational Progress (NAEP) by 10 points on a 500-point scale for the 2007 test of eighth graders, and by seven points for fourth graders.[19] The reading gap, furthermore, has been true longitudinally; the long-term trend version of the NAEP shows that boys have been behind girls since it was first administered in 1971.[20]

In writing, boys' scores are even worse. On a 300-point scale in this instance, the 2007 NAEP writing test gaps between males and females were 20 and 18 points in eighth and twelfth grades, respectively.[21] These gendered gaps, in a politically telling way, have been a featured element in the reporting of NAEP results, until recently preceding even race- and class-based gaps, which are actually more pronounced. The 2007 reading gap between white and black students was 27 points for fourth and eighth grades, and for those receiving free lunch and those not—a rough indicator of socioeconomic status—it was 29 and 25 points.[22] Privileging gender over—or equating it with—race and class is in line with a general absence of talk about the racial and classed dimensions of achievement among groups of boys. I return to this point shortly.

Boys are also more likely to drop out of school and repeat grades, and African American and Hispanic boys are disproportionately affected.[23] Boys suffer disproportionately from disciplinary measures, suspension, and expulsion from school, particularly boys of color.[24] Higher education access, too, has recently become an issue for males, with women outnumbering men in undergraduate enrollment at both two- and four-year colleges; women also made up 60 percent of graduate students in 2005, up by two percent from 2000.[25] These gaps are expected to increase through at least 2012.

Several social issues facing boys have also caused concern in the United States. Boys, for example, are disproportionately represented in special education.[26] While one could partly explain this by the underidentification of girls, many are worried that boys are targeted for their behavior problems more than their educational needs. This is particularly relevant to arguments over attention deficit disorder, which is disproportionately diagnosed in boys.[27] U.S. males are also more likely to commit suicide, abuse drugs and alcohol, and get caught up in the penal system.[28]

Many of these indicators have been true for many years; so, are men and masculinity really in *crisis* in the United States? The question is perhaps too subjective to answer, but some indicators do suggest men are not as economically stable as they once were (particularly men of color and/or working-class men). According to the National Center for Educational Statistics, in constant year 2000 U.S. dollars, the average man is making $5389 *less* than he was in 1971 when the statistics were first taken.[29] That's a 15 percent decrease in real dollars. In contrast, while the average woman still makes only US$1 to the average man's $1.39, women are earning 40 percent more than they did in 1971. Let me stress that women's gains have *in no way caused* men's declining average; it's more likely a function of better education for women and successful struggles for equal pay. Perceptions of a shaky future, however, are a reasonable response to men's own declines.

Like Australia, male *teacher* issues in the United States have also attracted publicity.[30] A National Education Association report shows that males make up only 21 percent of U.S. teachers.[31] Most males gravitate toward secondary school, however, leaving only 9 percent of elementary school teachers as male. The report framed the problem as a "40-year low" in the number of male teachers, a figure the media seized on and reported heavily. The media has given the causes of this limited analysis, however, largely focusing on economic and social decision-making: pundits blame the low pay and stereotypically female-appropriate nature of teaching.[32]

In spite of advocacy attention to social issues, the United States has seen less attention being paid to crucial questions asked elsewhere. Particularly absent is the question "Which boys?" are disadvantaged, acknowledging that not *all* boys are underachieving in schools. When the specter of boy crises appear, little is typically said about, say, gay or bisexual boys or boys with physical disabilities. The "backlash blockbusters" are focused—*because they don't say otherwise*[33]—on white middle-class boys, and they are largely consumed by white middle-class parents and teachers.

U.S. educators and scholars have not completely overlooked boys' diversities, though. In fact, several notable and influential works have shed light on the particular problems of African American boys and other boys of color.[34] Ferguson's *Bad Boys*, for example, shows how school disciplinary regimes construct black masculinity in harmful, entrenching ways. Importantly, Ferguson demonstrates that such constructions, including panics over black boys even in well-meaning circles, relegate African American boys to the equally damaging roles of "endangered species" or "criminal."

Constraints to the Growth of Boys' Education Issues

Given the varied concerns just cited—and returning to my foundational question—why have boys' educational issues not commanded more ubiquitous attention in U.S. media, educational research, and policy? Questions like this demand an ecological analysis to truly answer the issues, for the U.S. policy ecology is distinct from Australia's in important ways, ways that constrain the growth of boys' education issues.

Governmental and Educational Structures

The United States, like Australia, governs using a version of federalism, meaning that the federal and state governments have distinct areas of responsibility, guaranteed by the *U.S. Constitution*'s Tenth Amendment, which gives power to the states over anything not explicitly given to the Federal Government in the *Constitution*. Thus, waging war is a federal role, but education, because the *Constitution* doesn't mention it, is a state responsibility. This has interesting consequences. For example, the United States Government was able to mandate desegregation of schools in the 1950s as a civil rights issue (a federal role mandated by the Fourteenth Amendment), but it could not force Arkansas or Virginia to school black students and white students together; these states simply shut down their public schools rather than integrate them.

The United States' history of interventions in gender and education issues is inextricably bound up with such issues of federalism. Proponents advanced Title IX (discussed above) as a civil rights issue under the Fourteenth Amendment's mandate for "equal protection under the law." Though Title IX's scope is limited to those institutions accepting federal money—private schools and businesses are exempt—its provisions are broad, ensuring not only almost universal access to education (because the U.S. private school sector is small), but also limiting how single-sex programs, for instance, can be provided. Interventions targeted to boys, like the Detroit all-boys schools discussed in the next section, fall under Title IX rules, and this is part of what has prevented concerted boy-specific interventions until recently.

Federalism has also ensured the diffuse nature of boy concerns. Because no federal-level (de jure) policy is permissible—barring additional civil rights legislation for males, which would be unlikely—states are left to create their own policies for boys. Since movements for boys' education have been fractured and small, state-level policy has not been feasible, and, again, such policies run up against Title IX regulations. The only state, at this writing, to have attempted policy on boys' education has been Maine, and their process quickly turned from one on boys to one that considered gender broadly (boys *and* girls) due, some contended, to the political climate.[35]

A related constraint in the U.S. context not present in other nations is the frequent reliance of U.S. educational policy on judicial precedent.[36] In the United States' overtly litigious context, district courts and the Supreme Court arbitrate most important education policy regarding equity, civil rights, and religion in schools. *Brown v. Board of Education of Topeka, Kansas* (1954), for example, mandated U.S. schools desegregate by race, while *Engel v. Vitale* (1962) and others have prohibited school-sponsored prayer. This reliance on court precedence—where other nations largely develop and implement policy through bureaucracies—limits the ability for "equity" work for boys. Little or no precedent exists for considering males a disadvantaged group in U.S. law, and the effort, resources, and time involved in litigating for boys' educational issues have thus far been prohibitive.

Additionally, the U.S. private school sector is small relative to the public sector; approximately 6.1 million students attended private schools in 2003.[37] The vast majority of students—90 percent, or more than 53 million—attend public schools. The private sector's relative small size thus limits the exposure the U.S. populace has to boy-specific programs and single-sex schooling, and it limits the attention boys' issues have received relative to countries with larger private sectors—Australia and England, especially.

Social and Cultural Constraints

Another potential constraint on the growth of boys' issues is population size. While New Zealand has 4.2 million inhabitants, Australia 20.6 million, Canada 33.2 million, and the United Kingdom 60.9 million, the United States's population has reached nearly 303.8 million (all July 2008 estimates).[38] Capturing the attention of such a large population and getting them to act in a unified way presents no small challenge.

Other social and cultural factors, though, are likely more influential in containing boys' education discourses. First, the U.S. populace tends to think of "gender" issues as exclusively *women*'s issues. Despite movements that have increased awareness about masculinity and men's roles, particularly the 1990s' mythopoetic men's movements and the Promise Keepers movement among Christians, this remains true. Believing that gender is a women's issue limits receptivity to notions that boys might face problems *because they're male*. Similarly, girls' lingering disadvantages have yet to be solved, so focusing on boys can seem a distraction from, or threat to, more "traditional," accepted notions of gender equity.

Second, the U.S. focus for many decades has almost myopically been on school reform writ large. Since 1983, when the watershed *A Nation at Risk* report appeared, the major reform movements in the United States have concerned the setting of "standards" and the establishment of "accountability" and "choice" programs, typically centered on standardized testing.[39] After the passage of the *No Child Left Behind* (*NCLB*) Act in 2002, the federal law that mandated unprecedented testing and accountability measures, educators have focused even more on such reforms. With these large, looming issues in view, gender has taken a back seat even for girls.

Third, males' *gendered* concerns largely lack institutional infrastructures in the United States.[40] Good reasons exist for this, not the least of which is that social justice is aligned *against* most men's interests.[41] U.S. movements around men's issues have been fragmentary and local, often with little funding, few lobbyists, and thus scant impact on legislation. One possible exception is movements around divorce and custody support for men. Otherwise, no influential organization, like the National Organization for Women or the AAUW, exists to forward boys' educational issues.

Fourth and finally, movements toward boys' education tread a tense, ambivalent cultural ground in the United States due to their positions within larger, sometimes contradictory masculinity ideologies. Pollack's warnings in *Real Boys*, for example, about not shaming boys for attachments to their mothers fits well the "masculinity therapy" model.[42] Dobson, conversely, suggests fathers do what they can to force separation of a boy from his mother after the age of three; failure to do so risks turning boys into homosexuals, he claims.[43] The point is that arguments about boys must filter through these sometimes tense and contradictory cultural politics about masculinity in U.S. society. While some might lament the "poor boys" who suffer toxic gender roles, others perceive such lamentation as just another way to "soften" boys and to turn them away from "God's plan" for virile masculinity. On one hand, the diversity of views on masculinity could serve to increase the pool of those concerned about boys, for there is an explanation to suit many tastes. On the other hand, such ideological conflicts ensure the diffuse, fragmentary nature of the boy debates and limit the possibilities for unifying policies.

Boys' Education "On the Ground": Local Interventions

Earlier I argued that responsibility for education at the state and local level in the United States is a constraint on targeting boys' education issues in a concerted, coherent way. This is true at the federal and state levels in *explicit* policy. Local control of schools, however, along with growth in site-based management, where principals and staffs are responsible for most decision-making and policy, have facilitated growth in boys' education at the *local* level. Indeed, local interventions, *not policy*, are the chief sites for action around boys' education issues in the United States.

The mechanism explaining local growth of boys' education is, as in Australia, "de facto policy."[44] Individual classrooms and schools around the United States, largely invisible to the public, glean ideas from popular conservative texts. It is impossible to estimate how many educators have taken up such ideas or how much they have "cherry picked" and recontextualized such discourses to fit their needs.

Not all efforts at pedagogical and curricular interventions for U.S. boys have been under the radar or been based in conservative texts, though. In fact, several arguably more progressive initiatives have been created to address the unique and sometimes dire social and academic needs of African American boys.[45] Though other programs for boys are afoot, these African American-created and -centered programs stand out because they are sometimes heavily publicized, and they contradict notions that *all* boys are equally disadvantaged.

Perhaps the most visible example of Afrocentric programs for boys was the contentious creation of three all-boys, African American schools in Detroit, Michigan, in the early 1990s. A district court ruled, in 1991's *Garrett v. Board of Education of the School District of Detroit*, that the schools were in violation of girls' rights. Though advocates for the all-boys academies cited African American boys' unique educational needs, the plaintiffs argued the academies' establishment, because they were public schools, violated the Fourteenth Amendment's equal protection guarantee and Title IX, among others. While the all-boys schools were prohibited, chilling (until recently) the hopes of U.S. single-sex schooling advocates, the issues of boys—particularly African American boys—were put in the spotlight.

Other programs for African American boys have received less attention but demonstrate the possibilities and realities of local, nongovernmental programs. One such program I have previously worked with was the DREAMS program, formerly run by the Nehemiah Community Development Corporation of Madison, Wisconsin. This program, and others like it elsewhere, spawned from concerns for the lagging achievement of African American boys and the increasing social ills growing from this. Such programs often provide male mentors, particularly targeting boys without fathers at home, along with academic enrichment opportunities and (sometimes religious) counseling that are culturally relevant to the participants, integrated into local institutions, staffed by people who live and work in the communities and schools, and responsive to local needs.[46] The Paul Robeson Institute as described by Dance is another such program in Cambridge, Massachusetts.[47]

Caution is in order about such programs, despite whatever benefit and innovation they represent. First, the success of such programs, diffuse as they are and dependent upon the efforts of committed but often overworked community members, is never assured. Also, as Murtadah-Watts shows in her analysis of one Detroit academy intended to be all-male, discussed above, even programs *within* African American communities can unintentionally replicate racist regimes of control, particularly control over the bodies of black males.[48] Such findings suggest vigilance even when implementing purportedly "progressive" reform.

Contexts for Future Growth of Boys' Education Issues

While much of the debates and interventions surrounding boys' education have been constrained by governmental and social contexts and have been relegated to the local level, things are changing. Contexts are always already transforming. While results of changes are notoriously difficult to predict, there are noticeable trends in the United States bearing on the future of boys' education issues.

First, I mentioned above that testing and accountability movements limit the discourses of boys' education by monopolizing the public's attention. Contradictorily, though, these movements could well be a mainspring for *increased* growth in boys' education movements and policy. As Gillborn and Youdell have shown in England, a process of "triage" results when schools are judged on exams.[49] That is, those students who are "on the bubble" of passing the tests accrue the most funding and intervention, and those most likely to be "on the bubble" are boys. Given *NCLB*'s mandates to disaggregate test scores by sex, race, and class, the lagging performance of boys will be more apparent.[50] Considering also *NCLB*'s penalties for poor performance—or even lack of further progress for already well-performing schools—educators concerned for their jobs and their schools' survivals are forced to increasingly attend to poorly performing boys.

A second condition for future growth of boys' discourses is the increasing federal control over education. While, again, the federal education role is constitutionally limited (just as in Australia), new policy levers within *NCLB* are changing this.[51] Though federal funding accounts for only about 7 percent of total U.S. educational spending, the long-term underfunding of public schools makes even this amount indispensable, so *NCLB* holds sway over all educational authorities accepting federal monies. Similarly, the shift in federal funding from lump-sum distributions for states to targeted, competitive grants creates a situation in which the Federal Government can control what gets funded and what research corpuses and methods can be used. Thus, the government not only controls ideologies and create affinity groups, but it can also reward those supportive of its ideological positions. Because the state is a *deeply gendered* institution oriented to the benefit of already dominant males, this definitively advantages movements for boys' education.

Third, growth in boys' education discourses are encouraged by the current state of U.S. cultural politics.[52] In an era of conservative restoration and modernization with strong moves toward the Christian Right, groups critical of feminism and other forms of equity—sexuality, especially—are making concerted attacks.[53] Such groups, harkening in many instances to Muscular Christianity and other traditional gender beliefs, advocate boys' issues partly to "return" education to traditional, conservative Christian values existing before the impacts of feminism. Though I said earlier that cultural forces were working against boys' education in some ways, clearly

these forces are reconciling toward a rightist position, which ultimately helps boys' education.
A fourth context for growth of U.S. boys' education discourses is recent rollbacks of women's policy infrastructure, reasonably termed a backlash against women's gains. The Bush (the younger) administration, for example, within two months of assuming office in 2001, closed the White House Office of Women's Initiatives and Outreach, the office closest to the President's ear dedicated to women's and girls' concerns. Bush's fiscal year 2004 federal budget weakened the Women's Educational Equity Act by defunding implementation and support, and it slashed the budget of the Violence Against Women Program. In education, the Bush administration has twice commissioned "reviews" of Title IX, the first to clarify rules for university athletic programs, a reaction in part to high-profile cases of disbanding male athletic teams allegedly due to Title IX "restrictions." The second review targeted Title IX limitations on single-sex education, a change that took effect in late 2006. Considering the dwindling resources, infrastructure, and oversight for women's concerns, alongside multiple attempts to modify Title IX, women's groups are feeling besieged and conditions are ripe for increasing governmental programs for boys.

Fifth, boys' education panics and policies are finding fertile ground in mainstream media attention. *Newsweek*, a large-circulation weekly magazine, in January of 2006 ran a cover story proclaiming "The Boy Crisis," accompanied by a photograph of a group of scowling young toughs.[54] *Business Week*, a conservative-leaning national business magazine, earlier ran a cover proclaiming a "New Gender Gap" in which "boys are becoming the second sex," accompanied by an image of a huge girl, with arms folded and a cocky grin, looming over a reduced-size, downcast boy.[55] Of key concern, the continuing interest in boys' issues is markedly conservative, it generally lacks focused empirical study, and there is a dearth of high-profile U.S. voices of opposition—with the exception of scholars such as Michael Kimmel and David Sadker, high-profile reports by the think-tank Education Sector and by the AAUW, and a *Time* magazine cover story debunking "The Myth About Boys."[56] Combined with a perceptible decline in publicity of girls' remaining inequalities, the high profile of boys' issues has the potential to shift what the public and educators consider "real" gender issues.

Sixth, there are important public movements toward boys policy. Perhaps the most telling example was a high-profile civil rights complaint filed with the U.S. Department of Education in December 2005. Doug Anglin, a student at Milton High School in Massachusetts, and his lawyer father filed the claim, arguing that boys at his school are disadvantaged. As

Anglin told *The Boston Globe*, "From the elementary level, they [schools] establish a philosophy that if you sit down, follow orders, and listen to what they say, you'll do well and get good grades. Men naturally rebel against this."[57] Anglin also suggests, for example, that boys should be allowed to go to the bathroom whenever they want and that making boys decorate their notebooks is discriminatory. The complaint garnered local, national, and even international media coverage, with Internet sites of every ideology commenting on the story. (In line with the limitations discussed above, even conservative pundits, interestingly, were often dismissive of Anglin's claims of victimhood.) Most relevant to my argument, the complaint is clearly geared to trigger national policymaking. As *The Boston Globe* article says, "Anglin...said he brought the complaint in hope that the [U.S.] Education Department would issue national guidelines on how to boost boys' academic achievement." The Department of Education's Civil Rights Office dismissed Anglin's complaint in 2006, but clearly the pressure is on to bring federal policymaking to bear. It appears to be working.

Panic discourses are, in fact, beginning to be used by high-profile U.S. politicians. For example, in announcing publication of the 2004 Congressionally mandated biannual report on the education of girls and women,[58] the then-U.S. secretary of education, Rod Paige, said,

> It is clear that girls are taking education very seriously and that they have made tremendous strides,... The issue now is that boys seem to be falling behind. We need to spend some time researching the problem so that we can give boys the support to succeed academically.[59]

Paige's successor, Margaret Spellings, has also made public comment on boys' education concerns, being quoted in *Newsweek* as saying that boys' achievement gap "has profound implications for the economy, society, families and democracy."[60] Indeed, concern over boys has merited mention at the highest level of the United States government. In the 2005 State of the Union speech, annually the most watched and dissected political speech, George W. Bush voiced his concerns over boys:

> Now we need to focus on giving young people, especially young men in our cities, better options than apathy, or gangs, or jail. Tonight I propose a three-year initiative to help organizations keep young people out of gangs, and show young men an ideal of manhood that respects women and rejects violence. Taking on gang life will be one part of a broader outreach to at-risk youth, which involves parents and pastors, coaches and community leaders, in programs ranging from literacy to sports.

Though this speech doesn't directly address some core concerns circulating in popular-rhetorical literatures on boys' education, it represents a clear concern for masculinity crisis at top levels of the U.S. government. Though the United States has not yet gotten to Australia's level of policy attention on boys, it shows signs that it might soon. Thus, the United States might do well to learn from Australia's example in how to handle such potentially dangerous (in a social justice sense) and divisive reforms.

Cautions for Moving Forward

Given conditions favorable to those who would target educational resources to boys, both in Australia and the United States, I should note several cautions. First, the possibility that test scores could be the primary evidence for turning to boys' issues is problematic. Are tests valid? Do they measure what is truly of concern? While parents and educators are understandably concerned about achievement data, such indicators can elide social ills (risk taking, for example) and skill and attitude deficiencies (like citizenship attitudes) in boys, which deserve serious consideration. Test scores can also paint a more sanguine picture of *girls* than is actually the case.

Girls' needs may not always show up on a test, but they're still real. I join a chorus of those cautioning against the misguided conclusion that girls' needs have been solved. Girls still struggle with access to technology, high-status fields, and equitable social and economic outcomes from schooling. Serious problems remain in interpersonal relations girls have with boys. Many boys would benefit from antisexism, antiharassment, antiviolence, antihomophobia, and antiracism programs just as much as they would by literacy programs. Because boys' issues are relationally tied to girls' issues, these interventions serve the best interests of all students. Our ability to see indicators of need for such interventions, however, is weakened in a context of declining women's policy infrastructure, governmental devaluing of nonexperimental research, and rhetorical competition for "victim status."

Changes to Title IX in 2006 to allow single-sex education are also of particular concern. Though not advertised as such, changes to Title IX are necessary to clear hurdles for the public subsidizing of private single-sex schools through vouchers, a reform unambiguously supported by the Bush administration despite low public support. Indeed, like *NCLB* section 5131.(a).23's establishment of "innovative programs" for school choice including single-sex schools as well as sanctions elsewhere in the Act for anything less than achievement of standards by 100 percent of children (which no psychometrician will say is possible), most of *NCLB*'s provisions work toward a system friendly to vouchers and privatization.[61] Despite revisions to Title IX calling for "substantially equal" opportunities and the application of "evenhandedness" in program offerings, many current protections for both girls and boys are undercut and the ethic of "separate but equal," so hotly denounced in fights for African Americans' civil rights, is reinscribed on the basis of gender. This is not to say that single-sex schooling is inherently dangerous. It is not, though, a panacea; the U.S. Department of Education's own research says so.[62]

Possible racial implications, too, could follow from increasing attention to boys' issues. As has been shown in England, racism among white working-class boys has increased in conjunction with discourses of "white victimhood" following from the "underachieving boys" debate.[63] White victimhood discourses circulate in the United States too, so boy panics could potentially exacerbate racial tensions. Any focus on, or alternately exclusion of, African American boys could increase tensions, too, especially in view of the competitive nature characterizing boy debates thus far.

Again, each of these cautions is only that: a caution. Boys' education issues are neither inherently conservative nor inherently dangerous. Indeed, working with boys has tremendous potential for progressive, socially just education. Thus far, however, rightist movements and conservative authors have indeed co-opted these issues. Their conservative adherents don't, though, permanently taint such issues. With care and alertness, interventions can serve the interests of both boys and the society they help make up.

Conclusion

As this chapter has argued, boys' education issues in the United States—just as in Australia and other countries—are complex matters. Significant contextual differences exist between the U.S. reception of such matters and those of other countries. For the most part, movements concerned with boys' issues have had less "success" in the United States. Concern for boys has been diffuse, conservative in nature, locally contained, and structurally constrained. Yet, as this context evolves, boys' issues have high potential for growth. Comprehending this context requires an understanding of the history, governmental structure, and social and cultural peculiarities of U.S. education. It requires, in short, an understanding of those traits that shape and constrain what is possible, what is ordinary, and what is thinkable. Understanding these also suggests certain strategies dependent on the context. For the U.S. context, such strategies might include, among many others: targeting the local and diffuse sites of boys' education reform, usually individual schools and classrooms; working through the judicial system rather than, say, the bureaucracy, as one might in Australia; and working toward publicizing (pro)feminist messages in lay language and in easily accessible sources. In the end, with due caution sensitive to the ecology, any programs or interventions need not be conservative or destructive to already vulnerable groups. Rather, such programs *could* enrich us all. The Coda further explores what can be done to establish a progressive outcome for boys' education.

Coda
Hope and Strategy

To end this book, as the title of this coda implies, I will venture tentatively into two important topics: hope and strategy. It is imperative that progressive scholars, educators, and activists hold onto hope and develop strategy because the stakes couldn't be higher. Boys' education is a *central front* in what Apple has called "conservative modernization."[1] No other backlash discourse has been so successful in turning attention away from the focus of progressive and liberal movements. White victimhood may be common, but it is not mainstream.[2] The rich might talk about tax policy in terms of "class warfare," but common sense has not turned toward saying that the poor are really the advantaged in society. Boys' education discourses, though, have asserted victimhood and disadvantage as rallying cries, and the media and teachers have widely taken up these discourses. Conservative movements of all stripes have mobilized around boys' education, and here they have been more successful in turning back progressive gains and reestablishing the hegemony of males and patriarchy than anywhere else.

All is not lost, though. There is much to be hopeful about in boys' education and girls' education, and I believe progressive educators, activists, and scholars can derive some of that hope from exploring practice and policy that's progressive. I call this "situated strategizing," and it is applicable to nearly any issue and region, not just gender and not just the United States or Australia.

Hope

There has understandably been some despair over the state of gender equity in Australia, the United States, and elsewhere, as many have watched governments starve funding from previously robust programs for girls, and attention has quickly turned from girls to boys.[3] Some have claimed that, by the late 1990s, "gender reform [for girls] was on the wane almost everywhere."[4] Some have suggested, in fact, that Australia—and other contexts—may be seeing the "endgame" in policy and practice for girls.[5] Such assessments, though, are premature in my estimation. Instead, I suggest there are many reasons—*empirical* reasons—to have confidence that girls' gender equity will continue and prosper and that the worst impulses of boys' education movements can be overcome.

There have been many wise people in resistance movements who have articulated inspiration and reasons for hope better than I ever could. Yet, my purpose here has little to do with slogans that inspire, for indeed, those in the trenches of boy debates need much more to sustain real hope. I suggest that hope resides not in slogans but, even better, in the facts of the case. Simply put, there are concrete reasons to be hopeful. As Freire rightly asserts, "One of the tasks of the progressive educator, through a serious, correct political analysis, is to unveil opportunities for hope, no matter what the obstacles may be."[6] In this coda I want to "unveil opportunities for hope" about gender equity, despite the obstacles.

First, consider the political *advances* and progressive possibilities present in the shift to boys' education. With shifts in the economy and society that demand boys increase their communication skills and their ability to cooperate, the shift to trying to get boys to make such changes has many potential benefits for girls and women. There is also much progressive potential in understanding at a public level that boys, too, have a gender.[7] With such understandings, the violence, sexism, and unhealthy outcomes that are attached to certain versions of masculinity can be addressed and, one day perhaps, changed. This is being done in many schools, as can be seen in schools across the Boys' Education Lighthouse Schools (BELS) program, including those addressing violence and bullying.[8] The notion that *particular* boys are at risk is also being enshrined in practice, particularly with the inclusion of an entire module of the Success for Boys (SFB) program being dedicated to Indigenous boys.[9] In sum, Connell reminds us that there are numerous social justice goals for working with boys, including the goals of exposing boys to traditionally non-masculine knowledges, of improving their relationships with other humans, and of establishing justice in society.[10] "If we are not pursuing gender justice in the schools, then we are offering boys a degraded education—even though society may be offering them long-term privilege."[11] There is great reason to hope that attention to boys' education will pay off in a more socially just education for all.

A second reason for hope lies in the cyclical nature of panics in boys' education. As I pointed out in chapters 1 and 6, boy panics are not new. In fact, on a worldwide scale, boy panics have arisen continuously. Crotty shows that a crisis of masculinity was occurring in Australia at the turn of the twentieth century, just at the time that Forbush was also proclaiming a "boy problem" in the United States and G. Stanley Hall was proposing boys recapture manliness through playing "savages" in the classroom.[12] Boy panics also erupted in the middle of the twentieth century as concern was expressed over women teachers softening boys.[13] The recent reactionary movements around boys' issues, then, can be seen as a reiteration of a common pattern, not coincidentally occurring simultaneously with and in reaction to women's rights movements. Previous panics also have shown to subside and to give way to other, more fervent periods of (pro)feminist activity. I don't want to suggest that (pro)feminist activists simply "wait for it to pass," but there is comfort in knowing that regressive moments, historically, turn back. The sad part, of course, is that those regressive moments often sweep away many (but not all) of the fruits of previous hard work for social justice.

Third, though boys' education may be on the front burner currently, its success has not been monolithic or complete. Many teachers and other officials don't pay much (if any) attention to it or don't think much of it if they do. Indeed, many educators are sick of it and have "moved on." Many schools, in fact, have tried boys' education programs but have abandoned them for more equitable and effective means of teaching all students.[14] As with many other explosive, tense issues, teachers can quickly burn out their enthusiasm and can feel "overexposed" to the topics. Signs abound that this may be happening with boys' education.

Fourth, girls' educational policy and reform remain politically viable in Australia and elsewhere. State governments and professional organizations still produce girl-focused policies, particularly in science and technology. In terms of Australia, bear in mind that all of the state and territory governments' submissions to the Inquiry were markedly (pro)feminist in their orientation. Though this may be changing somewhat, elements of reasoned policymaking and even femocratic resistance remain within state departments of education. In the U.S. state of Maine, too (pro)feminists were able to similarly refocus the taskforce on boys' education to include the lingering needs of girls.[15]

There are (pro)feminist advocates remaining in politics too. Julia Gillard, formerly on the boys inquiry Committee and later in various ministerial posts (including at this writing being Australia's minister of education), said when I asked in 2003 if there were still possibilities for girls' education reform,

> Ah yeah, I think so, yes. I don't see this as necessarily ending the debate. I mean, if anything I think if you were able to point to issues for girls that still needed to be resolved.... All the evidence before the inquiry was that [course selection] was still very gendered in terms of girls not picking some of the hard sciences and those sorts of things, as well as boys not picking some of the you know so called "soft options."... Within the public policy debate those sorts of issues could still be raised and looked at... on their merits. I wouldn't be pessimistic that this has somehow closed down the debate on girls' education.

Even Brendan Nelson, former Committee chair and minister for education (and at this writing leader of the Opposition), sees girls' needs as a policy imperative, as he noted in a public hearing:

> There is absolutely nothing that will be recommended by us that will disadvantage girls. In fact, there are a couple of areas I have identified already in the course of this where I think girls [sic] outcomes could be improved. Career advice in education seems to be one.[16]

Though the careers advice package Nelson eventually put together included no provisions addressing gender, such statements, along with Gillard's, signal at least an awareness at the top levels of Australian government that girls have continuing needs.[17] Numerous U.S. politicians also continue to promote (pro)feminist stances, such as former first lady, senator, and presidential candidate Hillary Clinton.

At the same time, in Australia there seems to have been a slow down in the interest in boys' education initiatives at the Department of Education, Science and Training (DEST), particularly with the 2006 leaving of Brendan Nelson, a committed enthusiast, and the assumption of the portfolio by Julie Bishop. A great deal of money had been earmarked for boys' education from 2000 to 2005 (around A$30 million [US$26,175,000]; see table 4.3), but Bishop made no new promises of money beyond the SFB program, scheduled to end in 2008; this is a marked departure from the pace of previous planning within DEST (compare the timeline in the Appendix). By all accounts, Bishop wasn't as enthusiastic or committed to the boys' education policy as her predecessor. This may be cause for hope that other equity issues may be able to recapture some of this funding in the future. (Were the Coalition to regain control of government, thus making Nelson prime minister, this decreased attention to boys' education could change.)

Fourth, professional discretion among teachers continues apace. Though federal and state pressures may increasingly impinge upon teachers' daily practices, teachers still exercise critical thought when approaching conservative texts and policies. Springtown Religious School, described at length in chapter 5, provides a perfect example. Even though they frequently mentioned somewhat conservative theories and texts from which they drew, they cherry-picked those ideas that fit their needs, and these were often somewhat progressive.

Fifth, because of this continuing professional discretion and critical attitude among teachers, many schools continue to run programs focused on girls and women. As Apple and his colleagues have done in previous work, I point out such school-level work so that we might remind ourselves that progressive work does go on and so that we might "learn from each other, to combine our critical efforts."[18] One such example comes from the annual observation of International Women's Day at Gretna Green High School, part of the Riverside Schools Cooperative. Each year Ms. Gilbert runs it, and many activities go in it to target girls' needs:

> Well this year we had a breakfast; we started at 7:30 and as part of that we had four speakers. We had a [*inaudible*] liaison officer, we had a female police officer, they were all women speakers, we had the only female electrician at Queensland Rail and we had...the woman educator from the migrant women's education center, and they spoke....After that we had six workshops run by women for the girls, which went for an hour, which was just basically giving girls a go at something that they might not usually do.

Such programs are not without their problems, of course, including a lack of money and the grumblings of males at the school:[19]

> HIRST: I mean every year whenever we celebrate International Woman's Day you have this chorus of boys and men and male teachers around the school saying we need to get going on "International Men's Day." And the answer's always the same, "When you organize it," but they're not very good at organizing things for themselves.

Still, despite problems, many teachers remain actively committed to providing (pro)feminist programs. Funding sources for such girl-specific programs remain, too, despite the decline of government funding. For example, the Invergowrie Foundation is a grant-giving foundation that focuses on girls' education initiatives in Victoria, Australia. In the United States, the American Association of University Women and the National Science Foundation continue to grant a great deal of money for girls' education programming.

Several progressive programs also exist for *boys*. As I mentioned, many focus on violence and anti-bullying measures, but some schools also discuss issues like homophobia. Indeed, New Town High School, a BELS school in Hobart, Tasmania, runs a program called Pride and Prejudice, an antihomophobia program that reports dramatic successes in getting boys to discuss homophobia and to avoid the teasing, bullying, and violence that accompanies it. These and other programs provide hope that all isn't lost in making boys' education into a positive movement and to reengage the continuing needs of girls. With the knowledge of extant programs in schools, support (though admittedly mild) in the government, and historical precedent that suggests boys' crises are temporary, (pro)feminists can take *empirical* hope into the future.

Strategy

Hope does not materialize from wishing and waiting. Lingard, Hayes, Mills, and Christie suggest this with the following quote from Raymond Williams, from which they also take the subtitle of their book: "It is then in making hope practical, rather than despair convincing, that the ways to peace can be entered."[20] Making hope practical means utilizing strategies that bring the ideal into reality. In this last section, then, I humbly suggest strategies that educators and advocates might effectively employ in resisting conservative and regressive boys' education movements, both in contexts where debates have raged for years and also for those only beginning the discussion. Because every context is different, no single strategy will be effective for all situations. Instead, I suggest a concept of "situated strategizing" that relies on deeply understanding each policymaking arena.

To parse this "situated strategizing" concept somewhat, it relies on the notion that resistance to dominance—here, the dominance of recuperative masculinity politics—demands strategy and organization. This often becomes difficult for logistical and ideological reasons, but I mean organization in a loose sense. Strategy requires building alliances with those who share some, not necessarily all, of the goals progressives want to accomplish.[21] It requires, too, maintaining avenues for the distribution of progressive ideas that can reach and persuade the public. Such strategies are "situated" when they account for the specificities of place that interact with, change, subvert, block, shape, and transform the meanings of equity, social justice, gender, and education. They account for the many factors I have explored throughout this book, including a place's history and traditions; dynamics of race, class, gender, and sexuality; forms of government and the means of utilizing that government for policy change;

the economy; the social norms and cultural mores; previous policies that interact or limit change; and the key players that must be won over.

Strategy, firstly, demands building alliances, pulling in as many people as one can to help accomplish progressive goals. Rightist groups have had major successes in creating such strategic alliances, such as African American support of rightist voucher movements in the United States.[22] The same creation of coalitions can be done in progressive terms, as Apple shows in the "tactical alliances" formed to combat commercialism in schools.[23] Tensions, dangers, and missteps are endemic to the formulation of such alliances, but meaningful and effective progress requires building critical mass toward social change. Without such coalitions, (pro)feminist resistance to boys' education risks marginalization and a withering of needed support, finances, and infrastructure.

One way to build alliances is to use popular avenues of communication and to resist falling into solely esoteric language and thought as a means to communicate. As Martino and Berrill importantly remind us, "Those of us who are interested in teaching for social justice might do well to more consciously employ types of data which are compelling to those we wish to convince."[24] I might add that advocates should employ forms of *media* that are compelling, as well. As I described in chapter 1, one mainspring of the turn to boys' education, particularly true for Australia, has been its high visibility in the popular media. While some (pro)feminists have utilized these same venues, (pro)feminist arguments have largely been relegated to academic sources that are hard for the public to access. They're often hard to read too, for they're written in the register and vocabulary of scholarly argumentation. For widespread acceptance of (pro)feminist arguments, additional and more accessible forums and languages are needed. Books for parents, such as Paul Kivel's *Boys Will Be Men*, and practice-oriented books for teachers, such as Salisbury and Jackson's *Challenging Macho Values*, could serve as model resources.

On *situating* such strategies, I reiterate that the policy ecology metaphor is a key conceptualization for planning place-sensitive strategy. Such place-specificity isn't new to critical education, of course, but it is crucial that educators resisting policy also have a deep understanding of the peculiarities and specificities in which they work.[25] This requires identifying, at multiple levels (from the international to classrooms), those responsible for making decisions and policy; stakeholders and influential groups that must be resisted, answered, accommodated, or allied with; key arguments and ways to respond to them; driving forces of a policy; complicating factors and structural constraints; and strategic mistakes and open opportunities.

To bring it back to the specific contexts I have examined, how might strategy be formulated in Australia and the United States to retake ground in gender equity? While I don't have every answer, several tactics might be used as *part* of a strategy to hold off regressive boys' education movements and to continue to advance girls' needs.

The case studies in chapter 5, specifically how educators go about actually using boys' education policy, have important implications for progressive boys' education advocates. First, and perhaps most difficult, progressives must develop means of supporting innovative programs financially and logistically. It was clear in the case studies that a main motivator for taking on boys' education was obtaining grant funding that would ease the burdens of chronic economic scarcity. Grant funding attached to progressive programs might help forward progressive boys' and girls' education. Second, progressives must create and provide *accessible*, practice-oriented materials and research. Many schools and teachers seek out information on boys and how to teach them, and, since "backlash blockbusters" are the most accessible, they often turn to these.[26] If progressive, (pro)feminist work was more readily available, educators would have more nuanced, research-based choices. Third, (pro)feminists must create and provide progressive, *accessible* professional development. My own work has confirmed this for me. One woman who was attending a professional development workshop I conducted in 2006 for the Australasian Boys' Education Network announced that she came that day thinking that all boys were in crisis and doing badly in literacy, but that she would leave knowing that it was more complicated than that. "It's *not* all boys, is it?!" This replicates finding from the RSC case study where Martin Mills' professional development accessed teachers' lived knowledge and experience to suggest that a "which boys?" approach was most effective. Such professional development can be an important point of contact between progressives and practitioners.

Fourth, and finally, progressives would do well to create and disseminate interesting, entertaining curricular and extracurricular programming for use in schools. Central to the success of popular-rhetorical writers and consultants, again, is that they offer concrete plans (whether they work is another matter), including curriculum packets and after-school programming. Creating progressive materials and programs would help educators move beyond the typical programs prevalent today: have a star athlete come and read; father–son fishing trips. They might include more progressive programming like antisexist and antiviolence sports and physical education programs, utilize dance or outdoor education as anti-oppressive pedagogies, or incorporate progressive gender understandings into school counseling.[27]

At the level of winning the public debate, several opportunities also present themselves. Perhaps foremost, since so many people process gender and education issues through experiences with their own children (as discussed in chapter 3), targeting parenting as part of a situated strategizing has great potential. This includes the earlier suggestion for more accessible parenting texts, but it also requires using the public forum (like media and parent training) to *access* parents' knowledge of raising their own children as a way to think outside of regressive, conservative, or backlash notions about boys. Parents know *experientially* that their children are not simply stereotypes; sometimes they just need to be allowed to access that experience.

Similarly, some tweaking of the refrain "Which boys?" might be in order. For scholars doing research on boys, "which boys?" might be much more productively phrased as "which boys, when, where, with whom, and under what circumstances?" For practitioners, though, the question might become "What about *these* boys?", for some educators feel only paralysis when confronted by possibly having to fix oppression based on social class, race, region, or sexuality rather than simply having to read "boy-friendly" books in English class. Focusing on the boys in front of them is much less daunting.

Finally, I will ask a possibly heretical question to which many will, for good reason, object. Should progressive scholars begin reconceptualizing gender equity as not *primarily* an education issue? Many issues remain for schools regarding girls, of course, and I reject any notion that education should be abandoned as a feminist issue, but many of the most intractable problems for the advancement of women regard the *outcomes* of schooling, as reflected in much (pro)feminist research.[28] If advocates were to ask why girls are held back as women from utilizing their education for good jobs, equal pay, political representation, economic clout, and safety on the streets and at home, would the terms of the debate not change from the easily co-opted notion of test scores and suspensions toward the truly grinding disadvantages of poverty, miseducation, violence, and disrespect that our wives, mothers, sisters, and daughters experience daily? If girls' educations are so good, it might be asked, why do such things still plague societies?

Conclusion

I have, in this Coda, given empirical cause for hope for the (pro)feminist resistance to the regressive aspects of boys' education movements. (Pro) Feminists also have much to draw on for strategies that will hold off elements of the backlash against feminism and that will advance progressive education for all genders. A core element of such a strategy, I have argued, is utilizing a place-sensitive approach to analyzing policy contexts. In exploring the U.S. and Australian boys' education contexts, I hope I have demonstrated how a situated analysis can help us understand how and whether boys' education movements happen elsewhere. The comparison between the two countries gives insight into the complex factors that drive and change a policy ecology, highlighting how contextual differences and similarities make all the difference. Ultimately, I hope that this analysis provides a means for educators, parents, and citizens to resist backslides toward oppressive gender regimes and move, instead, toward a better education for *all* our children.

Appendix: Timeline of Australian Boys' Education Policy, 2000–2005

Date	Event
2000	
March 21	Education Minister Kemp gives the House Employment, Education, and Workplace Committee terms of reference for an Inquiry into the Education of Boys.
June 7 to 14	Inquiry and invitation for submissions is advertised in the media nationally.
October 5	Committee holds first public hearing in Canberra.
2001	
February 7	Dr. Nelson, chair of the Committee, leaves to become Minister of Education, Science and Training.
March 8	Elson takes over as Committee chair.
October 8	Gillard leaves the Committee to join shadow ministry.
November 10	Federal election held. Conservative Coalition maintains majority.
2002	
March 21	Committee changes names to Standing Committee on Education and Training after election and readopts the Inquiry. Bartlett becomes chair.
June 27	Committee holds final hearing, Canberra.
August 27	Cox leaves the Committee; Albanese joins the Committee.
October 21	Committee tables the report, *Boys: Getting It Right* (*BGIR*).
November 27	Minister Nelson convenes a boys' education forum in Canberra where he initiates the Boys' Education Lighthouse Schools (BELS) Programme.
2003	
March 3	Human Rights and Equal Opportunity Commission rejects the Catholic Education Office's plan to offer male-only scholarships.

Continued

Appendix Continued

Date	Event
March 27–29	Third biannual Boys to Fine Men (B2FM) Conference held in Newcastle, NSW, sponsored by the Department of Education, Science and Training (DEST).
April	DEST publishes a booklet aimed at parents and teachers, *Educating Boys: Issues and Information,* which outlines boys' education concerns, the arguments of the Inquiry, the need for more male teachers, and what the Commonwealth Government is doing to address the issue.
April 30	Minister Nelson announces the 226 schools for the first stage of BELS.
May 19	Minister Nelson requests a review of the Sex Discrimination Act (SDA) by the Attorney General with an eye to changing it to allow male-only scholarships.
June 26	Minister Nelson tables the Government's response to *BGIR, Boys' Education: Building on Successful Practice.*
November 5	GaiSheridan International awarded the *gender equity framework* recasting.
December	DEST releases the final report of BELS stage one, *Meeting the Challenge.*
2004	
March 10	Legislation is introduced in Parliament to amend Sex Discrimination Act to include exception for male-only teaching scholarships.
March 11	Minister Nelson opens consultation Web site, "Taking Schools to the Next Level" (now defunct; http://www.dest.gov.au/nef/schools), to give forum for parent and community input on educational issues, including questions on male teachers and boys' education.
March 19	An exemption to the SDA is given to the Catholic Education Office when they amend their request to provide teacher training scholarships to both females and males.
April 27	Minister Nelson announces the 38 stage-two BELS schools.
April 29	Draft of recast *gender equity framework* given to Ministerial Council on Education, Employment, Training and Youth Affairs.
May 3	Minister Nelson announces A$1 million male-only teacher training scholarship scheme.
May 11	The Senate Legal and Constitutional Legislation Committee releases a report on amendments to the SDA; the report suggests the bill proceed, but the opposition parties dissent.
June 10	Minister Nelson extends BELS stage-two grants to 13 additional schools at a cost of A$1.3 million.
June 22	Minister Nelson announces the Success for Boys (SFB) program, to cost A$19.4 million.

Continued

Appendix Continued

Date	Event
June 24	Minister Nelson announces a pilot study into father involvement.
October 9	Federal election. Coalition Government wins supermajority.
November 17	SDA amendment is introduced in Parliament for the third time.
2005	
February 11	National mapping of gender-related programs published (http://www.dest.gov.au/sectors/school_education/publications_resources/gender_specific_gender_related_curricula/default.htm [accessed January 30, 2008]).
April 3–5	Fourth biannual B2FM conference in Melbourne, sponsored by DEST.
June 27	Minister Nelson announces A$290,000 trial of SFB teaching materials and the schools that have won grants to trial them.
October 7	Applications for SFB stage one opens.
December 21	SFB stage one schools are announced.

NOTES

Foreword

1. Apple, *Educating the "Right" Way.*
2. Williams, *Key Words.*
3. See Apple, *State and the Politics; Educating the "Right" Way;* Apple and Buras, *Subaltern Speak.*
4. See, e.g., Dance, *Tough Fronts.*
5. Connell, *Masculinities;* Mills, *Challenging Violence;* Lingard, "Where to"; Lingard and Douglas, *Men Engaging Feminisms.*
6. Butler, *Undoing Gender.*
7. Dolby, Dimitriadis, and Willis, *Learning to Labor in New Times;* Willis, *Learning to Labor;* Weis, *Working Class without Work; Class Reunion.*
8. Johnson, "What Is Cultural Studies."
9. See also Du Gay et al., *Doing Cultural Studies.*
10. Apple, *State and the Politics.*
11. Apple, "Tasks."

Introduction

1. Bederman, *Manliness & Civilization,* chap. 3; Forbush, *Boy Problem;* Sexton, *Feminized Male.*
2. House of Representatives Standing Committee on Education and Training, *Boys: Getting It Right (BGIR).*
3. See Appadurai, *Modernity at Large.*
4. Titus, "Boy Trouble."
5. See Apple, *State and the Politics,* Introduction.
6. http://www.aph.gov.au/house/committee/edt/eofb/report.htm (accessed December 20, 2007).
7. Minister for Education Science and Training, *Boys' Education.*
8. For example, Lasswell, "Policy Orientation."
9. See deLeon, "Stages Approach"; Sabatier, *Theories.*
10. Ball, "Big Policies/Small World," 126.
11. Prunty, "Signposts"; see also Taylor et al., *Educational Policy.*
12. Weaver-Hightower, "Ecology Metaphor."
13. Ball, *Education Reform.*

14. Lingard, "Where to."
15. Apple, *Official Knowledge.*
16. I thank Jeffrey Sun for this question.
17. National Commission on Excellence in Education, *Nation at Risk.*
18. United Nations Children's Fund, *State,* 63.
19. For arguments about changing sex discrimination laws, see Magarey, "Sex Discrimination Amendment."
20. See also Mills, transcript of hearings for the Inquiry into the Education of Boys, *Hansard,* 651. Hereafter, all the hearing transcripts are referred to as "*Hansard,*" and the page numbers reflect the continuous pagination of the entire record.
21. Bourdieu, *Outline.*
22. For a fuller description of methods, see Weaver-Hightower, "Every Good Boy," chap. 2.

One: Gender and Education in a "New" Century: The "Boy Turn"

1. Weaver-Hightower, "Boy Turn."
2. Weaver-Hightower, "Crossing the Divide."
3. Organisation for Economic Co-operation and Development (OECD), *Education at a Glance.*
4. Connell, *Men and the Boys.*
5. See Biddulph, *Raising Boys;* Newkirk, "Misreading Masculinity"; Pollack, *Real Boys;* Sommers, *War against Boys;* Gurian, *Boys and Girls.*
6. Pirie, *Teenage Boys;* Newkirk, *Misreading Masculinity;* Martino, "Dickheads, Wuses, and Faggots"; Smith and Wilhelm, *Reading Don't Fix.*
7. For example, Newman, *Rampage;* Mills, *Challenging Violence;* Katz and Jhally, *Tough Guise.*
8. See Dance, *Tough Fronts;* Ferguson, *Bad Boys;* Sewell, *Black Masculinities and Schooling.*
9. Ferguson, *Bad Boys;* Murtadha-Watts, "Theorizing Urban Black Masculinity."
10. Newman, *Rampage.*
11. Espelage and Swearer, *Bullying in American Schools;* Martino and Pallotta-Chiarolli, *So What's a Boy?;* Rigby, *Meta-Evaluation;* Salisbury and Jackson, *Challenging Macho Values.* Recent works also focus on the ways, unique from boys, that girls bully and intimidate peers. See Wiseman, *Queen Bees and Wannabes;* Simmons, *Odd Girl Out.*
12. For example, Pollack, *Real Boys;* Kindlon and Thompson, *Raising Cain;* Cresswell, Rowe, and Withers, *Boys in School.*
13. Lingard et al., *Addressing the Educational Needs.*
14. Lillico, "The School Reforms Required."
15. Kipnis, *Angry Young Men;* Pollack, *Real Boys.* For an overview and critique of therapeutic approaches, see Gilbert and Gilbert, *Masculinity Goes to School,* 232–5; Lingard and Douglas, *Men Engaging Feminisms,* 40–42.
16. Jóhannesson, "To Teach"; Mills, Martino, and Lingard, "Attracting, Recruiting and Retaining"; NEA Research, *Status.*

17. See, e.g., Skelton, *Schooling the Boys*, chap. 6; Lesko, "Preparing to Coach"; Mills, "Issues."
18. Magarey, "Sex Discrimination Amendment"; Mills, Martino, and Lingard, "Attracting, Recruiting and Retaining." For reasons males avoid or leave teaching, see DeCorse and Vogtle, "In a Complex Voice"; King, *Uncommon Caring;* "Problem(s) of Men"; Sargent, *Real Men.*
19. Willis, *Learning to Labor;* Mac an Ghaill, *Making of Men;* Connell, *Masculinities;* Martino and Pallotta-Chiarolli, *So What's a Boy?.*
20. Ferguson, *Bad Boys;* Gillborn and Kirton, "White Heat"; Sewell, *Black Masculinities and Schooling;* Foley, *Learning Capitalist Culture.*
21. See, e.g., Meinhof, "Most Important Event"; Crotty, *Making the Australian Male;* Young, "Boy Talk."
22. See Bouchard, Boily, and Proulx, *School Success by Gender;* Epstein et al., *Failing Boys?.*
23. Martino and Berrill, "Boys, Schooling, and Masculinities."
24. Kenway and Willis, *Answering Back,* 49.
25. Biddulph, *Raising Boys.*
26. The "(pro)" in "(pro)feminist" refers almost exclusively to phenotypical or biological males who are allies of the feminist movement. This does not suggest, in my view, that men cannot be "feminists," but does acknowledge the political and theoretical complications of dominant groups identifying as oppressed groups. See Carr, "Same as It Never."
27. On military academies, see, e.g., Diamond, Kimmel, and Schroeder, "What's This."
28. Connell, *Men and the Boys.*
29. For mythopoetic understandings of masculinity, see, e.g., Bly, *Iron John;* Moore and Gillette, *King, Warrior, Magician, Lover;* Keen, *Fire in the Belly.*
30. Kenway and Willis, *Answering Back,* 61–62.
31. See, e.g., Faludi, *Backlash;* Arnot, David, and Weiner, *Closing the Gender Gap.*
32. Gillborn and Youdell, *Rationing Education,* chap. 6. For the U.S. context, see also McNeil, *Contradictions of School Reform.*
33. Lingard and Douglas, *Men Engaging Feminisms* (chap. 3).
34. Kenway and Willis, *Answering Back;* Lingard and Douglas, *Men Engaging Feminisms.*
35. For example, Arnot, David, and Weiner, *Closing the Gender Gap,* chaps. 6 and 8; Maynard, *Boys and Literacy,* 13–14.
36. Gee, Hull, and Lankshear, *New Work Order.*
37. Arnot, David, and Weiner, *Closing the Gender Gap,* 125–26.
38. Comaroff and Comaroff, "Millennial Capitalism," 307.
39. Yates, "Facts of the Case."
40. Apple's *Educating the "Right" Way,* 225–26, suggests all political projects consider both the good and bad senses of any ideology.
41. Apple and Oliver, "Becoming Right"; Weaver-Hightower, "Inventing."
42. Maguire and Ball, "Researching Politics," 270.
43. Yates, "Facts of the Case," 317.
44. Johnson, "What Is Cultural Studies."
45. United Nations Children's Fund (UNICEF), *State.*

46. See, e.g., Bennett, Emmison, and Frow, *Accounting for Tastes*, chap. 8. For the U.S. uses of international boy debates, see Titus, "Boy Trouble," 157.
47. See, e.g., *Hansard*, 1086–1114.
48. Name withheld, personal communication.
49. *Hansard*, 752–53.
50. Ministerial Council on Education Employment Training and Youth Affairs (MCEETYA) Gender Equity Taskforce, *Gender Equity*.
51. Sommers, *War against Boys*, chaps. 4 and 5.
52. Lingard and Douglas, *Men Engaging Feminisms;* Kenway and Willis, *Answering Back*.
53. Dowsett, "Masculinity, (Homo)Sexuality"; Connell, "Disruptions"; *Masculinities*.

Two: Masculinity "Down Under": The Roots of Boys' Education Policy in Australia

1. For critiques of the sometimes sloppy use of "hegemonic masculinity" see Connell, "Australian Masculinities"; Connell and Messerschmidt, "Hegemonic Masculinity."
2. Hollinsworth, *Race and Racism*.
3. Inglis, *Australian Colonists*, 53.
4. Crotty, *Making the Australian Male*.
5. Barcan, *History of Australian Education*.
6. Bernstein, *Class, Codes and Control (Volume 3)*.
7. *Hansard*, 1052.
8. Lingard, "Federalism in Schooling"; Lingard and Porter, *National Approach to Schooling*.
9. See http://www.assoa.nt.edu.au/history.html (accessed December 20, 2007).
10. Commonwealth Department of Education Science and Training, *Meeting the Challenge*.
11. See http://www.naa.gov.au/fsheets/fs195.html (accessed December 20, 2007).
12. Barcan, *History of Australian Education*.
13. National Office of Overseas Skills Recognition (NOOSR), *Country Education Profiles: Australia*.
14. See Thomson, *Schooling the Rustbelt Kids*.
15. Mills, *Hansard*, 651.
16. Ipswich Grammar School, "Ipswich Grammar School," prospectus, 2. http://www.igs.qld.edu.au/w04/media/UIL.pdf (accessed January 15, 2008). Ipswich Grammar isn't a pseudonym. Unlike the schools in chapter 5, I have not given pseudonyms to schools that advertise publicly. This also includes schools named in *BGIR* and other government publications, especially BELS reports (i.e., Commonwealth Department of Education Science and Training, *Meeting the Challenge;* Cuttance et al., *Boys' Education Lighthouse Schools*), where no expectation of privacy is given.

17. Cordwell, *Hansard,* 1122; Cook, *Hansard,* 609; Morgan, *Hansard,* 1121; Hopkins, *Hansard,* 1161.
18. Saunders and Evans, *Gender Relations in Australia;* Gerson and Peiss, "Boundaries, Negotiation, Consciousness."
19. Mills, *Challenging Violence*; Cunneen and Stubbs, "Fantasy Islands"; Evans, "Gun in the Oven."
20. Collins, Kenway, and McLeod, *Factors Influencing.*
21. See, e.g., Simms, "Two Steps Forward."
22. Lopez-Claros and Zahidi, *Women's Empowerment.*
23. Saunders and Evans, *Gender Relations in Australia,* chaps. 16–20.
24. Eisenstein, *Gender Shock; Inside Agitators;* Lingard and Douglas, *Men Engaging Feminisms;* Lingard, "Where to"; Marshall, "Policy Discourse Analysis"; Ailwood and Lingard, "Endgame"; van Acker, *Different Voices,* chap. 6.
25. Commonwealth Schools Commission, *National Policy*; Lingard, "Where to."
26. See, e.g., Coulter, "Boys Doing Good."
27. For example, Francis, *Hansard,* 134–46; submission to the Inquiry into the Education of Boys (hereafter, Submission) from the Eagle Forum.
28. Mills and Lingard, *Hansard,* 641–53; McLeod and Collins, *Hansard,* 153–64; submission by Australian Education Union.
29. *"Crocodile" Dundee,* directed by John Cornell.
30. Epstein and Johnson, *Schooling Sexualities,* 18.
31. Connell, *Masculinities.*
32. Ibid.
33. Goffman, "Arrangement between the Sexes."
34. *See* Henshall, *Hansard, 222;* Hewitson, *Hansard,* 805–6, 809; Horsell, *Hansard, 896.*
35. *BGIR,* xvii.
36. McKay et al., "Gender Equity," 233.
37. Ibid., 233.
38. Martino and Pallotta-Chiarolli*, So What's a Boy?, 249.*
39. Commonwealth Department of Education Science and Training, *Meeting the Challenge,* 83.
40. *Hansard, 83–84.*
41. Macintyre, *Concise History of Australia, 159.*
42. Crotty, *Making the Australian Male.*
43. See Damousi, "Marching to Different Drums"; Saunders and Bolton, "Girdled for War"; Macintyre, *Concise History of Australia,* chap. 7.
44. *Hansard, 1053–54.*
45. Connell, *Men and the Boys, 102.*
46. *The Block,* television series, Channel 9 (Australia), 2003.
47. *Australia: Beyond the Fatal Shore,* directed by Christopher Spencer, British Broadcasting Corporation, August 27, 2000; Dowsett, "Masculinity, (Homo) Sexuality"; Rowe and McKay, "A Man's Game."
48. Friend, "Choices, Not Closets"; Mac an Ghaill, "New Times"; "(In)Visibility"; Martino, "Dickheads, Wuses, and Faggots"; Martino and Pallotta-Chiarolli, *So What's a Boy?.*

49. *Hansard*, 228, 299, 636, 644, 645.
50. Connell, *Masculinities*.
51. Singleton et al., *Australian Political Institutions*.
52. Butts, *Assumptions Underlying Australian Education*.
53. Eisenstein, *Inside Agitators*.
54. Karmel, *Schools in Australia*.
55. Committee on Social Change and the Education of Women, *Girls, School & Society*.
56. Daws, "Quiet Achiever."
57. Commonwealth Schools Commission Working Party on the Education of Girls, *Girls and Tomorrow*.
58. Commonwealth Schools Commission, *National Policy*.
59. See Daws, "Quiet Achiever."
60. Australian Education Council, *National Action Plan*.
61. Ibid. See also Ailwood and Lingard, "Endgame"; Kenway and Willis, *Answering Back*.
62. Australian Education Council, *Listening to Girls*.
63. Daws, "Quiet Achiever," 102–3.
64. Ibid.
65. New South Wales Government Advisory Committee on Education Training and Tourism, *Challenges and Opportunities*.
66. Lingard, "Where to," 45.
67. MCEETYA Gender Equity Taskforce, *Gender Equity*.
68. Lingard, "Where to," 45.
69. Daws, "Quiet Achiever," 104.
70. *BGIR*, chapter 3, paragraphs 43–53.
71. *BGIR*, 68–69.
72. Lingard, "Where to"; Ailwood and Lingard, "Endgame"; Daws, "Quiet Achiever."
73. Epstein et al., "Schoolboy Frictions."
74. Compare punctuated equilibrium in Baumgartner and Jones, *Agendas and Instability*.
75. Apple, *Educating the "Right" Way*.
76. Smith, *Political Spectacle*.
77. For the conservative character of the debates, see Martino and Berrill, "Boys, Schooling, and Masculinities."
78. Latham, "Work, Family and Community: A Modern Australian Agenda," Australian Labor Party, http://www.alp.org.au//media/0204/20006891.html (accessed December 15, 2004).
79. Michael Bachelard and Rebecca DiGirolamo, "Latham Targets the Boy Crisis," *The Australian*, February 19, 2004.
80. Lingard, "Federalism in Schooling"; Lingard and Porter, *National Approach*.
81. Gillard later got significant attention as potentially the first female Leader of the Opposition, and she is, at this writing, Deputy Prime Minister, the highest office held by a woman in Australian history.
82. For example, for the U.K. context see Salisbury and Riddell, *Gender, Policy & Educational Change*.
83. Bouchard, Boily, and Proulx, *School Success by Gender*; Davison et al., "Boys and Underachievement"; Eyre, Lovell, and Smith, "Gender Equity Policy."

84. See http://www.maine.gov/education/gender_equity_taskforce (accessed December 30, 2007); Maine Task Force on Gender Equity in Education, *Final Report;* Madden, "Which Boys? Which Girls?"; Kevin Wack, "Thorny Politics Drive Panel Studying Gender," *Portland Press Herald,* March 29, 2006.

Three: *Boys: Getting It Right:* Inventing Boys through Policy

1. *Hansard,* 469.
2. NSW Government Advisory Committee on Education Training and Tourism, *Challenges and Opportunities.*
3. Edelman, *Constructing the Political Spectacle*; Smith, *Political Spectacle.*
4. I collected these exhibits in the Committee Secretariat's offices in Parliament House. Numerous exhibits were unavailable to me, on the Secretariat's discretion, including private, unpublished papers and correspondence. I did not copy various exhibits (or pages thereof) because their relevance to boys' education was unclear; uncopied pages were described in a log. In total, there were 38 available exhibits from which I copied no pages.
5. Lingard et al., *Addressing the Educational Needs;* Martino, Lingard, and Mills, "Issues in Boys' Education."
6. Jeff Singleton, staffer, personal correspondence.
7. *Hansard,* 899.
8. Apple, *Official Knowledge.*
9. Color blindness was discussed by Dennis Overton, Submission 195.
10. For a fuller description of my analysis methods, see Weaver-Hightower, "Every Good Boy," chapter 5.
11. This data was compiled from times listed by the *Hansard.* In cases where the end time was not recorded, I subtracted one minute from the recorded start time of the next witness(es).
12. *Hansard,* 415–27; exhibits 64 and 65.
13. Submission 165.
14. Submission 119.
15. Submission 150, 5, 21, 34.
16. Submission 75.
17. Fletcher, Submission 166.
18. Submission 103.
19. Ibid., 1.
20. Submission 117; hearings on October 5, 2000.
21. Submission 117, 16.
22. *Hansard,* 992.
23. Submission 117, 16.
24. Submission 168.
25. *Hansard,* 565–66.
26. For example, Browne and Fletcher, *Boys in Schools;* Fletcher, Hartman, and Browne, *Leadership in Boys' Education.*
27. *Hansard,* 1047.
28. Submission 166.
29. December 7, 2000, hearing.

30. *Hansard,* 513–14.
31. For example, Collins and Lea, *Learning Lessons.*
32. October 25, 2000, hearing; Submissions 111 and 111.1.
33. National Inquiry into the Teaching of Literacy, *Teaching Reading.*
34. May 8, 2002, hearing; Submission 188, 8.
35. Cadman, *Hansard,* 1234; May, *Hansard,* 1236.
36. Trent and Slade, *Declining Rates of Achievement.*
37. Sawford, *Hansard,* 876–77.
38. *Hansard,* 878.
39. *Hansard,* 1057.
40. See, e.g., The Endeavor Forum, submissions 21 through 21.4; *Hansard,* 134–46.
41. *Hansard,* 228.
42. *Hansard,* 230.
43. See, e.g., Sawford speaking to Ludowyke in *Hansard,* 24–26.
44. Apple, personal communication.
45. Hayes, "Mapping Transformations," 14.
46. Ibid.
47. Ibid.
48. Ibid.
49. Ibid.
50. Ibid., 15.
51. Examples of such research include Alloway et al., *Boys, Literacy and Schooling;* Lingard et al., *Addressing Educational Needs.* For more analysis of these, see Weaver-Hightower, "Every Good Boy," chap. 6.
52. *Hansard,* 1012.
53. *Hansard,* 309–10.
54. *Hansard,* 299.
55. Weaver-Hightower, "Boy Turn."
56. Heath, Submission 7, 1.
57. For example, Alloway and Gilbert, *Boys and Literacy;* Salisbury and Jackson, *Challenging Macho Values.*
58. For example, Browne and Fletcher, *Boys in Schools;* Brozo, *To Be a Boy;* Gurian, *Boys and Girls.*
59. Lingard, "Where to."
60. *Hansard,* 230–31.
61. On reasons males might avoid teaching, see King, *Uncommon Caring;* Sargent, *Real Men.*
62. *Hansard,* 1049–50.
63. See also Woodside-Jiron, "Critical Policy Analysis," 534.
64. Apple, *Educating the "Right" Way.*
65. I thank Martin Mills for suggesting this point.
66. *Hansard,* 716.
67. Lingard et al., *Addressing the Educational Needs;* Martino, Lingard, and Mills, "Issues in Boys' Education."
68. Kenway and Willis, *Answering Back.*
69. For other reasons, see King, *Uncommon Caring;* Sargent, *Real Men.*

70. For example, Pidgeon, "Learning Reading"; Brozo, *To Be a Boy;* Millard, *Differently Literate;* Martino, "It's Not the Way"; "Dickheads, Wuses, and Faggots."
71. Pressley, *Reading Instruction That Works.*
72. See Gee, "It's Theories"; Erickson, "Arts, Humanities, and Sciences"; Lincoln and Cannella, "Dangerous Discourses"; Lather, "This *IS* Your Father's."
73. Sommers, *War against Boys.*
74. Ibid. See also Gurian and Stevens, *Minds of Boys;* James, *Teaching the Male Brain.*
75. For exceptions, see Von Drehle, "Boys Are All Right"; Mead, *Truth About Boys and Girls.*
76. Edelman, *Constructing the Political Spectacle;* Smith, *Political Spectacle.*
77. Apple, *Educating the "Right" Way;* the title also makes a nod to Mills, Martino, and Lingard, "Getting Boys' Education 'Right'."

Four: Means to an End: The Resulting Initiatives

1. Smith, *Political Spectacle,* 6.
2. Ball, *Education Reform,* 15.
3. Smith, *Political Spectacle,* 6.
4. Williams, *Marxism and Literature,* 132–34.
5. Apple, *Educating the "Right" Way,* 225–26.
6. In British and Australian usage, "practise" is the verb form and "practice" is a noun.
7. Ministerial media release, October 21, 2002, http://www.dest.gov.au/ministers/nelson/oct02/n216_211002.htm (accessed January 2, 2007).
8. Belinda Tyrrel, DEST, personal correspondence.
9. Apple and Weis, "Ideology and Practice."
10. http://www.pco.com.au/boys2005/Indigenous%20Program%20at%20a%20glance%20NEW.pdf (accessed September 9, 2007).
11. McLean, "Boys and Education"; Mills, "Issues in Implementing."
12. Hartman and Fletcher, 2003 conference abstracts, session C, 16–17.
13. Fletcher, Submission 166.
14. Compare to Connell, "Teaching the Boys."
15. For example, Mayer, "What is the Place"; Gee, "It's Theories"; Walker, "After Methods, Then What?"; Lincoln and Cannella, "Dangerous Discourses"; Lather, "This *IS* Your Father's."
16. Williams, *Marxism and Literature.*
17. For the various traditions, see Weaver-Hightower, "Boy Turn."
18. Daws, "Quiet Achiever."
19. MCEETYA, *The Adelaide Declaration.*
20. DEST request for quotes, 2.
21. Ibid.
22. Ibid.
23. Weaver-Hightower, "Boy Turn"; "Every Good Boy."
24. Commonwealth Department of Education Science and Training, *Annual Report 2003–04;* actual cost from http://www.dest.gov.au/NR/

rdonlyres/16DB0588-5B78-4728-968D-C65BD29CECA2/4057/8apr05_compiled_murray_order.pdf (accessed March 1, 2007).
25. See, for example, Hartman, *I Can Hardly Wait;* Browne and Fletcher, *Boys in Schools;* Fletcher, Hartman, and Browne, *Leadership in Boys' Education.*
26. GaiSheridan International, "New Perspectives."
27. GaiSheridan International, *"Gender Equity Framework,"* draft dated April 29, 2004, 3.
28. Helen McDevitt, DEST, personal communication. Gai Sheridan, personal communication.
29. GaiSheridan International, *"Gender Equity Framework."*
30. Connell, *Masculinities.*
31. For example, American Association of University Women, *Hostile Hallways.*
32. Deem, "Gendered Governance," suggests that, in the United Kingdom, local control by lay people creates numerous gendered effects, so local control may be largely geared to the interests of men.
33. Lingard, "Where to."
34. GaiSheridan International, *"Gender Equity Framework,"* 18–19.
35. See also Weaver-Hightower, "Dare the School."
36. For example, Connell, "Disruptions"; "Teaching the Boys"; Jackson and Salisbury, "Why Should Secondary Schools"; Denborough, "Step by Step"; Kenway and Willis, *Answering Back.*
37. GaiSheridan International, *"Gender Equity Framework,"* 15.
38. On boys' policy as an endgame for girls' policy, see Ailwood, "A National Approach"; Ailwood and Lingard, "Endgame."
39. Erica Cervini, "Male Teaching 'Bribes' Under Fire," *The Age,* March 15, 2003. http://www.theage.com.au/articles/2003/03/15/1047583739672.html (accessed January 15, 2008).
40. Ibid.
41. Ministerial media release, May 19, 2003.
42. Ministerial media release, May 4, 2004.
43. http://www.hreoc.gov.au/sex_discrimination/exemption/decision.html (accessed November 7, 2005).
44. Australian Senate Legal and Constitutional Legislation Committee, *Provisions.*
45. Magarey, *Sex Discrimination Amendment.*
46. Latham, "Work, Family and Community: A Modern Australian Agenda," Australian Labor Party, http://www.alp.org.au//media/0204/20006891.html (accessed December 15, 2004).
47. Quoted in Magarey, *Sex Discrimination Amendment,* 7.
48. Ibid.
49. Work produced for DEST includes Alloway et al., *Boys, Literacy and Schooling;* Collins, Kenway, and McLeod, *Factors Influencing;* Lingard et al., *Addressing the Educational Needs;* Trent and Slade, *Declining Rates of Achievement;* Munns et al., *Motivation and Engagement.*
50. See http://www.standards.dfes.gov.uk/beaconschools (accessed January 7, 2008).
51. Commonwealth Department of Education Science and Training, *Meeting the Challenge; Meeting the Challenge (Summary Report).*

52. Cuttance et al., *Boys' Education Lighthouse Schools*.
53. Commonwealth Department of Education Science and Training, *Meeting the Challenge*.
54. Ibid., 135–38.
55. Sommers, *War against Boys*.
56. http://www.boyslighthouse.edu.au/faqs.htm (accessed July 25, 2003).
57. According to Peter Cuttance (personal communication), the director of BELS phase two, his team performed a blind review on each application. The scores were then sent to DEST for final school selections, a process that was not blind. He confirms that the selection was done in a way to balance state and territory representation, government and nongovernment, rural and urban.
58. National Office of Overseas Skills Recognition (NOOSR), *Country Education Profiles: Australia*.
59. Commonwealth Department of Education Science and Training, *Meeting the Challenge*.
60. Ibid., 3.
61. Ibid., 19, 79–80.
62. Ibid., 30–31.
63. http://www.nqsf.edu.au/nqsf/schoolproject (accessed November 13, 2005).
64. Cuttance et al., *Boys' Education Lighthouse Schools*.
65. Commonwealth Department of Education Science and Training, "Success for Boys," brochure.
66. The modules can be viewed at http://www.successforboys.edu.au/boys/sfb_download_modules,15732.html (accessed January 7, 2008).
67. Lingard, "Where to."
68. Mills, "Shaping the Boys' Agenda."
69. Marceau, "Steering from a Distance"; Kickert, "Steering at a Distance."
70. See, e.g., McCracken, "Surviving Shock and Awe."
71. http://www.dest.gov.au/sectors/school_education/policy_initiatives_reviews/key_issues/boys_education/default.htm (accessed January 7, 2008).
72. McCracken, "Surviving Shock and Awe."
73. Gee, "Identity as Analytic Lens," 100.
74. Apple and Pedroni, "Conservative Alliance Building."
75. Lingard et al., *Addressing the Educational Needs*.
76. Weaver-Hightower, "Dare the School."
77. Alan Finder, "Giving Disorganized Boys the Tools for Success," *The New York Times*, January 1, 2008.
78. Weaver-Hightower, "Inventing."
79. http://www.thegitd.com/Subscription%20Form%20for%20newsletter.pdf (accessed January 7, 2008); http://www.understandingfamilies.com/parenting_dvds.html (accessed January 7, 2008); the fee for the summer institute is for the "early-bird" rate.
80. http://www.thegitd.com/requirements.htm (accessed January 7, 2008).
81. http://www.thegitd.com/Website%20Grant%20document.doc (accessed January 7, 2008).
82. Red River Valley Writing Project workshop, November 3, 2007.

83. Bouchard, Boily, and Proulx, *School Success by Gender;* Martino and Kehler, "Male Teachers."
84. For the context of Iceland, see Jóhannesson, "To Teach." For the United States, see, e.g., NEA Research, *Status.*

Five: "Getting It Right" in the Schoolhouse: Two Case Studies

1. Sheridan, Street, and Bloome, *Writing Ourselves.*
2. Other helpful case studies of implementing boys' education policy include Maynard, *Boys and Literacy,* and Rowan et al., *Boys, Literacies and Schooling.* Viewing these alongside my cases give a broader view of the diversity of boys' education program implementation.
3. Fairclough, *Critical Discourse Analysis,* chap. 6.
4. See Connell, "Cool Guys"; *Masculinities;* Martino, "Dickheads, Wuses, and Faggots."
5. He particularly referenced Slade's *Listening to the Boys.*
6. Moir and Jessel, *Brain Sex.*
7. Commonwealth Department of Education Science and Training, *Meeting the Challenge.*
8. Lingard and Douglas, *Men Engaging Feminisms.*
9. See Thomson, *Schooling the Rustbelt Kids,* chap. 5.
10. For the other major "little things" done by SRS on boys' education, see Weaver-Hightower, "Every Good Boy," chap. 7.
11. Rowan et al., *Boys, Literacies and Schooling;* Kenway and Willis, *Answering Back.*
12. Howe and Strauss, *Millennials Rising.*
13. Rowan et al., *Boys, Literacies and Schooling.*
14. http://www.hillbrook.qld.edu.au/groups/parentresources/page.cfm (accessed March 3, 2007). The page has since been removed.
15. Weaver-Hightower, "Every Good Boy," chap. 7.
16. See Cahill, "On Teachers' Access."
17. Bernstein, *Class, Codes and Control (Volume 3).*
18. Gillborn and Youdell, *Rationing Education.*
19. Ball, *Education Reform,* chap. 2.
20. Lingard, "Where to."
21. Ball, *Education Reform,* chap. 2.
22. Apple and Oliver, "Becoming Right"; Apple, *Educating the "Right" Way;* Weaver-Hightower, "Inventing."
23. Kindlon and Thompson, *Raising Cain.*
24. Eisenhart and Finkel, *Women's Science,* chap. 7.
25. Gurian, *Fine Young Man;* Pollack, *Real Boys;* Sax, *Why Gender Matters;* Sommers, *War against Boys.*
26. Frederick Kunkle, "Boy 'Tribes' On Frontier in Reading," *The Washington Post,* January 8, 2005.
27. Costello, "Teaching Chivalry."
28. See, e.g., Newkirk, "Brain Research"; Titus, "Boy Trouble," 153–55.

Six: Boys' Education in the United States: What Australia's Example Tells Us

1. See, e.g., Bleach, *Raising Boys' Achievement;* Noble and Bradford, *Getting It Right;* Head, *Understanding the Boys.*
2. See, e.g., Arnot and Gubb, *Adding Value.*
3. Bouchard, Boily, and Proulx, *School Success by Gender.* See also Frank et al., "Tangle of Trouble," 120. On New Zealand, see Ross, "Nats Call for Inquiry"; Rebecca Fox, "Boys—the Classroom Timebomb," *New Zealand Herald,* May 20, 2006, http://www.nzherald.co.nz/section/1/print.cfm?c_id=1&objectid=10382701&pnum=0 (accessed January 20, 2008).
4. Jóhannesson, "To Teach."
5. Weaver-Hightower, "Boy Turn."
6. Weaver-Hightower, "Ecology Metaphor."
7. Clarke, *Sex in Education,* part four.
8. Bederman, *Manliness & Civilization.*
9. Sexton, *Feminized Male;* Sommers, *War against Boys.*
10. See, e.g., Sadker and Sadker, *Failing at Fairness,* 214–15; Kimmel, *Manhood in America;* Kidd, *Making American Boys.* See Crotty, *Making the Australian Male,* for a discussion of these in Australian educational history.
11. For example, Women on Words and Images, *Dick and Jane;* Frazier and Sadker, *Sexism in School.*
12. National Commission on Excellence in Education, *Nation at Risk.*
13. American Association of University Women, *Hostile Hallways; Growing Smart;* Orenstein, *School Girls;* Pipher, *Reviving Ophelia;* Sadker and Sadker, *Failing at Fairness.*
14. Martino and Berrill, "Boys, Schooling, and Masculinities"; Mills, "Shaping the Boys' Agenda."
15. See also Sommers, *Who Stole Feminism?*
16. See, e.g., Brozo, *To Be a Boy;* Gurian and Stevens, *Minds of Boys;* James, *Teaching the Male Brain.*
17. See, however, Lesko, *Masculinities at School;* Kimmel, *Manhood in America;* Thorne, *Gender Play;* Ferguson, *Bad Boys;* Brown, *Black Superheroes.*
18. Lingard, "Where to."
19. National Center for Education Statistics, *Nation's Report Card: Reading,* 30, 12.
20. National Center for Educational Statistics, *NAEP 2004 Trends.*
21. National Center for Educational Statistics, *Nation's Report Card: Writing,* 12, 40.
22. National Center for Education Statistics, *Nation's Report Card: Reading,* 11, 29, 13, 31.
23. Bae et al., *Trends in Educational Equity,* 34.
24. Ferguson, *Bad Boys.*
25. National Center for Educational Statistics, *Condition of Education,* http://nces.ed.gov/programs/coe/2007/section1/indicator09.asp (accessed January 15, 2008).
26. See, e.g., Oswald et al., "Trends."
27. Pollack, *Real Boys,* 253–56.

28. http://www.cdc.gov/ncipc/dvp/Suicide/SuicideDataSheet.pdf (accessed January 15, 2008); http://www.drugabusestatistics.samhsa.gov/NSDUH/2k6NSDUH/tabs/Sect2peTabs1to42.htm#Tab2.42B (accessed January 1, 2008); http://www.drugabusestatistics.samhsa.gov/NSDUH/2k6NSDUH/tabs/Sect1peTabs1to46.htm (accessed January 15, 2008); http://www.ojp.gov/bjs/glance/cpgendpt.htm (accessed January 15, 2008).
29. National Center for Educational Statistics, *Condition of Education 2004*, 133.
30. Mills, Martino, and Lingard, "Attracting, Recruiting and Retaining."
31. NEA Research, *Status*.
32. See, e.g., Rowden-Racette, "Endangered Species."
33. Morrison, *Playing in the Dark*, 72.
34. See also Lopez, *Hopeful Girls*.
35. Maine Task Force on Gender Equity in Education, *Draft Report;* Kevin Wack, "Thorny Politics Drive Panel Studying Gender," *Portland Press Herald*, March 29, 2006.
36. I am indebted to Fazal Rizvi for this insight.
37. http://nces.ed.gov/programs/digest/d06/tables/dt06_055.asp (accessed January 12, 2008).
38. *CIA World Factbook* online, https://www.cia.gov/library/publications/the-world-factbook/fields/2119.html (accessed June 23, 2008).
39. National Commission on Excellence in Education, *A Nation at Risk;* Smith, *Political Spectacle;* McNeil, *Contradictions*.
40. My reference here is to infrastructure with *explicit* charter to serve men's interests as gendered actors. Compelling arguments can be made that the vast majority of institutions *implicitly* serve the interests of men.
41. See Connell, *Masculinities*, chap. 10.
42. Ibid. See also Kenway and Willis, *Answering Back;* Lingard and Douglas, *Men Engaging Feminisms*.
43. Dobson, *Bringing up Boys*, 122, quoting J. Nicolosi.
44. Lingard, "Where to."
45. See, e.g., Smith, "Black Boys."
46. Ladson-Billings, *The Dreamkeepers*.
47. Dance, *Tough Fronts*.
48. Murtadha-Watts, "Theorizing Urban Black Masculinity."
49. Gillborn and Youdell, *Rationing Education*, chap. 6.
50. *NCLB* does not have sex/gender as part of its Adequate Yearly Progress measure, the most direct link to accountability punishments. It does, however, in Title I, part A, subpart 1, section 1111, (b)(3)(C)(xiii) require that states disaggregate test scores by gender and, in (h)(1)(C)(i), that states report these disaggregated numbers to the secretary of education and the public.
51. See Smith, *Political Spectacle*, 239–41.
52. I thank Michael Apple for this point.
53. Apple, *Official Knowledge; Educating the "Right" Way*.
54. Tyre, "Trouble with Boys."
55. Conlin, "New Gender Gap."
56. See, e.g., Kimmel, "What About the Boys?"; Sadker, "Educator's Primer"; Mead, *Truth about Boys and Girls;* AAUW, *Where the Girls Are;* Von Drehle, "Boys Are All Right."

57. Tracy Jan, "Schoolboy's Bias Suit: Argues System Is Favoring Girls," *The Boston Globe*, January 26, 2006.
58. Freeman, *Trends in Educational Equity*.
59. http://www.ed.gov/news/pressreleases/2004/11/11192004b.html (accessed November 19, 2004).
60. Tyre, "Trouble with Boys."
61. See Smith, *Political Spectacle*, 234–35.
62. U.S. Department of Education, *Single-Sex Versus Coeducational*.
63. Gillborn and Kirton, "White Heat."

Coda: Hope and Strategy

1. Apple, *Educating the "Right" Way*.
2. Gillborn and Kirton, "White Heat."
3. See, e.g., Blackmore and Sachs, "Progression and Regression."
4. Kenway and Willis, *Answering Back*, 210.
5. Ailwood, "National Approach"; Ailwood and Lingard, "Endgame"; Lingard, "Where to."
6. Freire, *Pedagogy of Hope*, 2.
7. Mills, *Challenging Violence*.
8. Commonwealth Department of Education Science and Training, *Meeting the Challenge*, 13, 23, 74, 113.
9. Herbert and Gilbert, *Success for Boys*.
10. Connell, "Teaching the Boys," 221–24.
11. Ibid., 224.
12. Crotty, *Making the Australian Male;* Forbush, *Boy Problem;* Bederman, *Manliness & Civilization*.
13. Sexton, *Feminized Male*.
14. Lingard et al., *Addressing the Educational Needs*.
15. Kevin Wack, "Thorny Politics Drive Panel Studying Gender," *Portland Press Herald*, March 29, 2006.
16. *Hansard*, 296.
17. Ministerial press release, April 16, 2004.
18. Apple, *State and the Politics*, 24. See also Apple and Beane, *Democratic Schools*.
19. See also Kenway and Willis, *Answering Back*.
20. Lingard et al., *Leading Learning*, 1.
21. See Apple, *Official Knowledge*, chap. 5, on temporary alliances.
22. Apple and Oliver, "Becoming Right"; Apple and Pedroni, "Conservative Alliance Building."
23. Apple, *Educating the "Right" Way*, 250–53.
24. Martino and Berrill, "Boys, Schooling, and Masculinities," 113.
25. See, e.g., Gruenewald, "Best of Both Worlds"; Edmondson, *Prairie Town*.
26. Mills, "Shaping the Boys' Agenda."
27. Mills, *Challenging Violence;* Gard, "Being Someone Else"; Forbes, *Boyz 2 Buddhas*.
28. See especially Collins, Kenway, and McLeod, *Factors Influencing*.

Bibliography

Ailwood, Jo. "A National Approach to Gender Equity Policy in Australia: Another Ending, Another Opening?" *International Journal of Inclusive Education* 7, no. 1 (2003): 19–32.

Ailwood, Jo, and Robert Lingard. "The Endgame for National Girls' Schooling Policies in Australia?" *Australian Journal of Education* 45, no. 1 (2001): 9–22.

Alloway, Nola, and Pam Gilbert, eds. *Boys and Literacy—Teaching Units*. Carlton, Australia: Curriculum Corporation, 1997.

Alloway, Nola, Peter Freebody, Pam Gilbert, and Sandy Muspratt. *Boys, Literacy and Schooling: Expanding the Repertoires of Practice*. Canberra, Australia: Commonwealth Department of Education, Science and Training, 2002.

American Association of University Women. *Growing Smart: What's Working for Girls in School*. Washington, DC: AAUW, 1995.

———. *Hostile Hallways: The AAUW Survey on Sexual Harassment in America's Schools*. Washington, DC: AAUW, 1993.

———. *How Schools Shortchange Girls*. New York: Marlowe & Company, 1992.

———. *Where the Girls Are: The Facts about Gender Equity in Education*. Washington, DC: AAUW, 2008.

Appadurai, Arjun. *Modernity at Large: Cultural Dimensions of Globalization*. Minneapolis: University of Minnesota Press, 1996.

Apple, Michael W. *Educating the "Right" Way: Markets, Standards, God, and Inequality*. 2nd ed. New York: Routledge, 2006.

———. *Official Knowledge: Democratic Education in a Conservative Age*. 2nd ed. New York: Routledge, 2000.

———, ed. *The State and the Politics of Knowledge*. New York: RoutledgeFalmer, 2003.

———. "The Tasks of the Critical Scholar/Activist in Education." In *Methods at the Margins*, edited by R. Winkle-Wagner, D. H. Ortloff, and C. Hunter. New York: Palgrave, forthcoming.

Apple, Michael W., and James A. Beane, eds. *Democratic Schools*. Alexandria, VA: Association for Supervision and Curriculum Development, 1995.

Apple, Michael W., and Kristen L. Buras, eds. *The Subaltern Speak: Curriculum, Power, and Educational Struggles*. New York: Routledge, 2006.

Apple, Michael W., and Anita Oliver. "Becoming Right: Education and the Formation of Conservative Movements." *Teachers College Record* 97 (1996): 419–45.

Apple, Michael W., and Thomas C. Pedroni. "Conservative Alliance Building and African American Support of Vouchers: The End of *Brown's* Promise or a New Beginning?" *Teachers College Record* 107, no. 9 (2005): 2068–105.

Apple, Michael W., and Lois Weis. "Ideology and Practice in Schooling: A Political and Conceptual Introduction." In *Ideology and Practice in Schooling*, edited by Michael W. Apple and Lois Weis, 3–33. Philadelphia: Temple University Press, 1983.

Arnot, Madeleine, Miriam David, and Gaby Weiner. *Closing the Gender Gap: Postwar Education and Social Change*. Cambridge, England: Polity Press, 1999.

Arnot, Madeleine, and Jennifer Gubb. *Adding Value to Boys' and Girls' Education: A Gender and Achievement Project in West Sussex*. West Sussex, England: West Sussex County Council, 2001.

Australian Bureau of Statistics. "Population Distribution – 2001 (Map)." http://www.abs.gov.au/Ausstats/abs@.nsf/Lookup/361F400BCE3AB8ACCA256CAE00053FA4 (accessed January 15, 2008).

Australian Education Council. *Listening to Girls: A Report of the Consultancy Undertaken for the Australian Education Council Committee to Review the National Policy for the Education of Girls*. Melbourne, Australia: Curriculum Corporation, 1992.

———. *National Action Plan for the Education of Girls 1993–1997*. Carlton, Victoria, Australia: Curriculum Corporation, 1993.

Australian Senate Legal and Constitutional Legislation Committee. *Provisions of the Sex Discrimination Amendment (Teaching Profession) Bill 2004*. Canberra: Australian Senate, 2004.

Bae, Yupin, Susan Choy, Claire Geddes, Jennifer Sable, and Thomas Snyder. *Trends in Educational Equity of Girls and Wome*n. Washington, DC: U.S. Department of Education, National Center for Education Statistics, 2000.

Ball, Stephen J. "Big Policies/Small World: An Introduction to International Perspectives in Education Policy." *Comparative Education* 34, no. 2 (1998): 119–30.

———. *Education Reform: A Critical and Post-Structural Approach*. Buckingham, England: Open University Press, 1994.

———. "Researching inside the State: Issues in the Interpretation of Elite Interviews." In *Researching Education Policy: Ethical and Methodological Issues*, edited by David Halpin and Barry Troyna, 107–20. London: Falmer Press, 1994.

Barcan, Alan. *A History of Australian Education*. Melbourne, Australia: Oxford University Press, 1980.

Baumgartner, Frank R., and Bryan D. Jones. *Agendas and Instability in American Politics*. Chicago: University of Chicago Press, 1993.

Bederman, Gail. *Manliness & Civilization: A Cultural History of Gender and Race in the United States, 1880–1917*. Chicago: University of Chicago Press, 1995.

Bennett, Tony, Michael Emmison, and John Frow. *Accounting for Tastes: Australian Everyday Cultures*. Cambridge, England: Cambridge University Press, 1999.

Bernstein, Basil. *Class, Codes and Control (Volume 3: Towards a Theory of Educational Transmissions)*. Rev. ed. London: Routledge & Kegan Paul, 1977.

Biddulph, Steve. *Raising Boys: Why Boys Are Different—and How to Help Them Become Happy and Well-Balanced Men.* Sydney: Finch, 1998.
Blackmore, Jill, and Judyth Sachs. "Progression and Regression: Managing Diversity, Equity and Equal Opportunity." In *Performing and Reforming Leaders: Gender, Educational Restructuring, and Organizational Change,* edited by Jill Blackmore and Judyth Sachs, 221–44. Albany: State University of New York Press, 2007.
Bleach, Kevan, ed. *Raising Boys' Achievement in Schools.* Staffordshire, England: Trentham Books, 1998.
Bly, Robert. *Iron John: A Book About Men.* Reading, MA: Addison-Wesley, 1990.
Bouchard, Pierette, Isabelle Boily, and Marie-Claude Proulx. "School Success by Gender: A Catalyst for Masculinist Discourse." Ottawa, Ontario, Canada: Status of Women Canada, 2003.
Bourdieu, Pierre. *Outline of a Theory of Practice.* Translated by Richard Nice. Cambridge, England: Cambridge University Press, 1977.
Brown, Jeffrey A. *Black Superheroes, Milestone Comics, and Their Fans.* Jackson: University Press of Mississippi, 2001.
Browne, Rollo, and Richard Fletcher, eds. *Boys in Schools: Addressing the Real Issues—Behaviour, Values, and Relationships.* Sydney: Finch, 1995.
Brozo, William G. *To Be a Boy, to Be a Reader: Engaging Teen and Preteen Boys in Active Literacy.* Newark, DE: International Reading Association, 2002.
Bryson, Bill. *Down Under.* London: Doubleday, 2000.
Butler, Judith. *Undoing Gender.* New York: Routledge, 2004.
Butts, R. Freeman. *Assumptions Underlying Australian Education.* New York: John Wiley, 1955.
Cahill, William. "On Teachers' Access to the Discourse of Education." *Teachers College Record* (2005), http://www.tcrecord.org/content.asp?contentid=12210 (accessed January 15, 2008).
Carr, Brian. "Same as It Never Was: Masculinity and Identification in Feminism." In *Masculinities at School,* edited by Nancy Lesko, 323–42. Thousand Oaks, CA: Sage, 2000.
Clarke, Edward H. *Sex in Education, or, a Fair Chance for Girls.* Boston: Houghton Mifflin, 1873.
Collins, Bob, and Tess Lea. *Learning Lessons: An Independent Review of Indigenous Education in the Northern Territory.* Darwin, Australia: Northern Territory Department of Education, 1999.
Collins, Cherry, Jane Kenway, and Julie McLeod. *Factors Influencing the Educational Performance of Males and Females in School and Their Initial Destinations after Leaving School.* Canberra, Australia: Commonwealth Department of Education, Training, and Youth Affairs, 2000.
Comaroff, Jean, and John L. Comaroff. "Millennial Capitalism: First Thoughts on a Second Coming." *Public Culture* 12, no. 2 (2000): 291–343.
Committee on Social Change and the Education of Women. *Girls, School & Society: Report by a Study Group to the Schools Commission.* Canberra: Australian Schools Commission, 1975.
Commonwealth Department of Education, Science and Training. *Annual Report 2001–02.* Canberra, Australia: Department of Education, Science and Training, 2002.

Commonwealth Department of Education, Science and Training. *Annual Report 2002–03.* Canberra, Australia: Department of Education, Science and Training, 2003.

———. *Annual Report 2003–04.* Canberra, Australia: Department of Education, Science and Training, 2004.

———. *Annual Report 2004–05.* Canberra, Australia: Department of Education, Science and Training, 2005.

———. "Educating Boys: Issues and Information." Brochure. Canberra, Australia: Department of Education, Science and Training, 2003. http://www.dest.gov.au/sectors/school_education/publications_resources/profiles/educating_boys.htm (accessed January 30, 2008).

———. *Meeting the Challenge: Guiding Principles for Success from the Boys' Education Lighthouse Schools Programme Stage One 2003.* Canberra: Australian Government Department of Education, Science and Training, 2003.

———. *Meeting the Challenge: Guiding Principles for Success from the Boys' Education Lighthouse Schools Programme Stage One 2003 (Summary Report).* Canberraa: Australian Government Department of Education, Science and Training, 2003.

Commonwealth Department of Education, Training and Youth Affairs. *Annual Report 2000–01.* Canberra, Australia: Department of Education, Training and Youth Affairs, 2001.

Commonwealth Schools Commission Working Party on the Education of Girls. *Girls and Tomorrow: The Challenge for Schools.* Canberra, Australia: Commonwealth Schools Commission, 1984.

Commonwealth Schools Commission. *The National Policy for the Education of Girls in Australian Schools.* Canberra, Australia: Commonwealth Schools Commission, 1987.

Conlin, Michelle. "The New Gender Gap." *Business Week*, May 26, 2003.

Connell, R. W. "Australian Masculinities." In *Male Trouble: Looking at Australian Masculinities*, edited by Stephen Tomsen and Mike Donaldson, 9–21. Melbourne, Australia: Pluto Press, 2003.

———. "Cool Guys, Swots and Wimps: The Interplay of Masculinity and Education." *Oxford Review of Education* 15, no. 3 (1989): 291–303.

———. "Disruptions: Improper Masculinities and Schooling." In *Beyond Silenced Voices: Class, Race, and Gender in United States Schools*, edited by Lois Weis and Michelle Fine, 191–208. Albany: State University of New York Press, 1993.

———. *Masculinities.* 2nd ed. Berkeley: University of California Press, 2005.

———. *The Men and the Boys.* Berkeley: University of California Press, 2000.

———. "Teaching the Boys: New Research on Masculinity and Gender Strategies for Schools." *Teachers College Record* 98, no. 2 (1996): 206–35.

Connell, R. W., and James W. Messerschmidt. "Hegemonic Masculinity: Rethinking the Concept." *Gender & Society* 19, no. 6 (2005): 829–59.

Cornell, John. *"Crocodile" Dundee*, DVD. United States: Paramount Pictures, 1986.

Costello, Judith. "Teaching Chivalry in the Classroom." *Teacher Magazine*, October 23, 2007. http://www.teachermagazine.org/tm/articles/2007/10/23/07knights_web.h19.html (accessed January 16, 2008).

Coulter, Rebecca Priegert. "Boys Doing Good: Young Men and Gender Equity." *Educational Review* 55, no. 2 (2003): 135–45.
Cresswell, John, Kenneth J. Rowe, and Graeme Withers. *Boys in School and Society*. Camberwell, Victoria, Australia: ACER Press, 2002.
Crotty, Martin. *Making the Australian Male: Middle-Class Masculinity 1870–1920*. Melbourne, Australia: Melbourne University Press, 2001.
Cunneen, Chris, and Julie Stubbs. "Fantasy Islands: Desire, 'Race' And Violence." In *Male Trouble: Looking at Australian Masculinities*, edited by Stephen Tomsen and Mike Donaldson, 69–90. Melbourne, Australia: Pluto Press, 2003.
Cuttance, Peter, Wesley Imms, Sally Godhino, Elizabeth Hartnell-Young, Jean Thompson, Keryn McGuinness, and Gregory Neal. *Boys' Education Lighthouse Schools Stage Two Final Report 2006*. Canberra, Australia: Department of Education, Science and Training, 2007.
Damousi, Joy. "Marching to Different Drums: Women's Mobilisations 1914–1939." In *Gender Relations in Australia: Domination and Negotiation*, edited by Kay Saunders and Raymond Evans, 350–75. Sydney: Harcourt Brace, 1992.
Dance, L. Janelle. *Tough Fronts: The Impact of Street Culture on Schooling*. New York: RoutledgeFalmer, 2002.
Davison, Kevin G., Trudy A. Lovell, Blye W. Frank, and Ann B. Vibert. "Boys and Underachievement in the Canadian Context: No Proof for Panic." In *The Politics of Gender and Education: Critical Perspectives*, edited by Suki Ali, Shereen Benjamin, and Malanie L. Mauthner, 50–63. Houndsmill, England: Palgrave Macmillan, 2004.
Daws, Leonie. "The Quiet Achiever: *The National Policy for the Education of Girls*." In *A National Approach to Schooling in Australia? Essays on the Development of National Policies in Schools Education*, edited by Bob Lingard and Paige Porter, 95–110. Canberra: Australian College of Education, 1997.
DeCorse, Cynthia J. Benton, and Stephen P. Vogtle. "In a Complex Voice: The Contradictions of Male Elementary Teachers' Career Choice and Professional Identity." *Journal of Teacher Education* 48, no. 1 (1997): 37–46.
Deem, Rosemary. "Gendered Governance: Education Reform and Lay Involvement in the Local Management of Schools." In *Gender, Policy & Educational Change: Shifting Agendas in the UK and Europe*, edited by Jane Salisbury and Sheila Riddell, 191–207. London: Routledge, 2000.
deLeon, Peter. "The Stages Approach to the Policy Process: What Has It Done? Where Is It Going?" In *Theories of the Policy Process*, edited by Paul A. Sabatier, 19–32. Boulder, CO: Westview Press, 1999.
Denborough, David. "Step by Step: Developing Respectful and Effective Ways of Working with Young Men to Reduce Violence." In *Men's Ways of Being*, edited by Christopher McLean, Maggie Carey, and Cheryl White, 91–115. Boulder, CO: Westview Press, 1996.
Diamond, Diane, Michael S. Kimmel, and Kirby Schroeder. "'What's This About a Few Good Men?': Negotiating Gender in Military Education." In *Masculinities at School*, edited by Nancy Lesko, 231–49. Thousand Oaks, CA: Sage, 2000.
Dobson, James. *Bringing up Boys*. Wheaton, IL: Tyndale House Publishers, 2001.
Dolby, Nadine, Greg Dimitriadis, and Paul Willis. *Learning to Labor in New Times*. New York: RoutledgeFalmer, 2004.

Dowsett, Gary W. "Masculinity, (Homo)Sexuality and Contemporary Sexual Politics." In *Male Trouble: Looking at Australian Masculinities*, edited by Stephen Tomsen and Mike Donaldson, 22–39. Melbourne, Australia: Pluto Press, 2003.

Du Gay, Paul, Stuart Hall, Linda Janes, Hugh Mackay, and Keith Negus. *Doing Cultural Studies: The Story of the Sony Walkman.* Thousand Oaks, CA: Sage, 1997.

Edelman, Murray. *Constructing the Political Spectacle.* Chicago: University of Chicago Press, 1988.

Edmondson, Jacqueline. *Prairie Town: Redefining Rural Life in the Age of Globalization.* Lanham, MD: Rowman & Littlefield, 2003.

Eisenhart, Margaret A., and Elizabeth Finkel. *Women's Science: Learning and Succeeding from the Margins.* Chicago: University of Chicago Press, 1998.

Eisenstein, Hester. *Gender Shock: Practicing Feminism on Two Continents.* Boston: Beacon Press, 1991.

———. *Inside Agitators: Australian Femocrats and the State.* Philadelphia: Temple University Press, 1996.

Epstein, Debbie, and Richard Johnson. *Schooling Sexualities.* Buckingham, England: Open University Press, 1998.

Epstein, Debbie, Jannette Elwood, Valerie Hey, and Janet Maw, eds. *Failing Boys? Issues in Gender and Achievement.* Buckingham, England: Open University Press, 1998.

———. "Schoolboy Frictions: Feminism and 'Failing' Boys." In *Failing Boys?: Issues in Gender and Achievement*, edited by Debbie Epstein, Jannette Elwood, Victoria Hey, and Janet Maw, 3–18. Buckingham, England: Open University Press, 1998.

Erickson, Frederick. "Arts, Humanities, and Sciences in Educational Research and Social Engineering in Federal Education Policy." *Teachers College Record* 107, no. 1 (2005): 4–9.

Espelage, Dorothy L., and Susan M. Swearer, eds. *Bullying in American Schools: A Social-Ecological Perspective on Prevention and Intervention.* Mahwah, NJ: Lawrence Erlbaum, 2004.

Evans, Raymond. "A Gun in the Oven: Masculinism and Gendered Violence." In *Gender Relations in Australia: Domination and Negotiation*, edited by Kay Saunders and Raymond Evans, 197–218. Sydney: Harcourt Brace, 1992.

Eyre, Linda, Trudy A. Lovell, and Carrol Ann Smith. "Gender Equity Policy and Education: Reporting on/from Canada." In *The Politics of Gender and Education: Critical Perspectives*, edited by Suki Ali, Shereen Benjamin, and Malanie L. Mauthner, 67–86. Houndsmill, England: Palgrave Macmillan, 2004.

Fairclough, Norman. *Critical Discourse Analysis: The Critical Study of Language.* London: Longman, 1995.

Faludi, Susan. *Backlash: The Undeclared War against American Women.* New York: Anchor Books, 1991.

———. *Stiffed: The Betrayal of the American Man.* New York: William Morrow, 1999.

Ferguson, Ann Arnett. *Bad Boys: Public Schools in the Making of Black Masculinity.* Ann Arbor: University of Michigan Press, 2000.

Fletcher, Richard, Deborah Hartman, and Rollo Browne, eds. *Leadership in Boys' Education*. Newcastle, Australia: University of Newcastle, 1999.
Foley, Douglas E. *Learning Capitalist Culture: Deep in the Heart of Tejas*. Philadelphia: University of Pennsylvania Press, 1990.
Forbes, David. *Boyz 2 Buddhas: Counselling Urban High School Male Athletes in the Zone*. New York: Peter Lang, 2004.
Forbush, William Byron. *The Boy Problem*. Boston: Pilgrim Press, 1901.
Frank, Blye, Michael Kehler, Trudy Lovell, and Kevin Davison. "A Tangle of Trouble: Boys, Masculinity and Schooling—Future Directions." *Educational Review* 55, no. 2 (2003): 119–33.
Frazier, Nancy, and Myra Sadker. *Sexism in School and Society*. New York: Harper & Row, 1973.
Freeman, Catherine E. *Trends in Educational Equity for Girls and Women: 2004*. Washington, DC: U.S. Department of Education, 2004.
Freire, Paulo. *Pedagogy of Hope: Reliving Pedagogy of the Oppressed*. Translated by Robert R. Barr. London: Continuum, 1992.
Friend, Richard A. "Choices, Not Closets: Heterosexism and Homophobia in Schools." In *Beyond Silenced Voices: Class, Race, and Gender in United States Schools*, edited by Lois Weis and Michelle Fine, 209–35. Albany: State University of New York Press, 1993.
GaiSheridan International. "*Gender Equity Framework* for Schools 2004." Unpublished draft [dated April 29, 2004] of gender equity policy for the Australian Ministerial Council for Education, Employment, Training and Youth Affairs, 2004.
———. "New Perspectives on Gender Equity in Australian Schools." Unpublished report on gender equity recasting for the Australian Ministerial Council for Education, Employment, Training and Youth Affairs, 2004.
Gard, Michael. "Being Someone Else: Using Dance in Anti-Oppressive Teaching." *Educational Review* 55, no. 2 (2003): 211–23.
Gee, James Paul. "Identity as an Analytic Lens for Research in Education." *Review of Research in Education* 25 (2000–2001): 99–125.
———. "It's Theories All the Way Down: A Response to *Scientific Research in Education*." *Teachers College Record* 107, no. 1 (2005): 10–18.
Gee, James Paul, Glynda A. Hull, and Colin Lankshear. *The New Work Order: Behind the Language of the New Capitalism*. Boulder, CO: Westview Press, 1996.
Gerson, Judith M., and Kathy Peiss. "Boundaries, Negotiation, Consciousness: Reconceptualizing Gender Relations." *Social Problems* 32, no. 4 (1985): 317–31.
Gilbert, Rob, and Pam Gilbert. *Masculinity Goes to School*. London: Routledge, 1998.
Gillborn, David, and Alison Kirton. "White Heat: Racism, Under-achievement and White Working-Class Boys." *International Journal of Inclusive Education* 4, no. 4 (2000): 271–88.
Gillborn, David, and Deborah Youdell. *Rationing Education: Policy, Practice, Reform, Equity*. Buckingham, England: Open University Press, 2000.
Gilligan, Carol. *In a Different Voice: Psychological Theory and Women's Development*. Cambridge, MA: Harvard University Press, 1982.

Goffman, Erving. "The Arrangement between the Sexes." *Theory and Society* 4, no. 3 (1977): 301–31.
Gruenewald, David A. "The Best of Both Worlds: A Critical Pedagogy of Place." *Educational Researcher* 32, no. 4 (2003): 3–12.
Gurian, Michael. *Boys and Girls Learn Differently! A Guide for Teachers and Parents.* San Francisco: Jossey-Bass, 2001.
———. *A Fine Young Man: What Parents, Mentors, and Educators Can Do to Shape Adolescent Boys into Exceptional Men.* New York: Jeremy P. Tarcher/Putnam, 1998.
Gurian, Michael, and Kathy Stevens. *The Minds of Boys: Saving Our Sons from Falling Behind in School and Life.* San Francisco: Jossey-Bass, 2005.
Hartman, Deborah, ed. *Educating Boys: The Good News: Insights from a Selection of Papers Presented at the 4th Biennial Working with Boys, Building Fine Men Conference.* Callaghan, NSW, Australia: University of Newcastle, 2006.
———. *I Can Hardly Wait Till Monday: Women Teachers Talk About What Works for Them and for Boys.* Newcastle, Australia: University of Newcastle, 1999.
Hayes, Debra. "Mapping Transformations in Educational Subjectivities: Working within and against Discourse." *International Journal of Inclusive Education* 7, no. 1 (2003): 7–18.
Head, John. *Understanding the Boys: Issues of Behaviour and Achievement.* London: Falmer Press, 1999.
Herbert, Jeannie, and Rob Gilbert. *Success for Boys: Indigenous Boys Module.* Carlton, Australia: Curriculum Corporation, 2006.
Hollinsworth, David. *Race and Racism in Australia.* 2nd ed. Katoomba, Australia: Social Science Press, 1998.
House of Representatives Standing Committee on Education and Training. *Boys: Getting It Right. Report on the Inquiry into the Education of Boys.* Canberra: Parliament of the Commonwealth of Australia, 2002.
Howe, Neil, and William Strauss. *Millennials Rising: The Next Great Generation.* New York: Vintage, 2000.
Inglis, Kenneth Stanley. *The Australian Colonists: An Exploration of Social History, 1788–1870.* Melbourne, Australia: Melbourne University Press, 1974.
Jackson, David, and Jonathan Salisbury. "Why Should Secondary Schools Take Working with Boys Seriously?" *Gender and Education* 8, no. 1 (1996): 103–15.
James, Abigail Norfleet. *Teaching the Male Brain: How Boys Think, Feel, and Learn in School.* Thousand Oaks, CA: Corwin Press, 2007.
Jóhannesson, Ingólfur Ásgeir. "To Teach Boys and Girls: A Pro-Feminist Perspective on the Boys' Debate in Iceland." *Educational Review* 56, no. 1 (2004): 33–42.
Johnson, Richard. "What Is Cultural Studies Anyway?" *Social Text*, no. 16 (1986–1987): 38–80.
Karmel, Peter H. *Schools in Australia: Report of the Interim Committee for the Australian Schools Commission.* Canberra: Australian Schools Commission, 1973.
Katz, Jackson, and Sut Jhally. *Tough Guise: Violence, Media, and the Crisis in Masculinity.* DVD. Directed by Sut Jhally. Northampton, MA: Media Education Foundation, 1999.

Keen, Sam. *Fire in the Belly: On Being a Man.* New York: Bantam Books, 1991.
Kenway, Jane, and Sue Willis. *Answering Back: Girls, Boys and Feminism in Schools.* London: Routledge, 1998.
Kickert, Walter. "Steering at a Distance: A New Paradigm of Public Governance in Dutch Higher Education." *Governance: An International Journal of Policy and Administration* 8, no. 1 (1995): 135–57.
Kidd, Kenneth B. *Making American Boys: Boyology and the Feral Tale.* Minneapolis: University of Minnesota Press, 2004.
Kimmel, Michael S. *Manhood in America: A Cultural History.* New York: Free Press, 1996.
———. "What About the Boys?" *WEEA Digest*, November 2000.
Kindlon, Dan, and Michael Thompson. *Raising Cain: Protecting the Emotional Life of Boys.* New York: Ballantine Books, 2000.
King, James R. "The Problem(s) of Men in Early Education." In *Masculinities at School*, edited by Nancy Lesko, 3–26. Thousand Oaks, CA: Sage, 2000.
———. *Uncommon Caring: Learning from Men Who Teach Young Children.* New York: Teachers College Press, 1998.
Kipnis, Aaron. *Angry Young Men: How Parents, Teachers, and Counselors Can Help "Bad Boys" Become Good Men.* San Francisco: Jossey-Bass, 1999.
Kivel, Paul. *Boys Will Be Men: Raising Our Sons for Courage, Caring, and Community.* Gabriola Island, BC, Canada: New Society Publishers, 1999.
Ladson-Billings, Gloria. *The Dreamkeepers: Successful Teachers of African American Children.* San Francisco: Jossey-Bass, 1994.
Lasswell, Harold D. "The Policy Orientation." In *The Policy Sciences*, edited by Daniel Lerner and Harold D. Lasswell. Stanford, CA: Stanford University Press, 1951.
Latham, Mark. "Work, Family and Community: A Modern Australian Agenda." Australian Labor Party. http://www.alp.org.au//media/0204/20006891.html (accessed March 15, 2005).
Lather, Patti. "This *IS* Your Father's Paradigm: Government Intrusion and the Case of Qualitative Research in Education." *Qualitative Inquiry* 10, no. 1 (2004): 15–34.
Lesko, Nancy, ed. *Masculinities at School.* Thousand Oaks, CA: Sage, 2000.
———. "Preparing to *Teach* Coach: Tracking the Gendered Relations of Dominance, on and off the Football Field." In *Masculinities at School*, edited by Nancy Lesko, 187–212. Thousand Oaks, CA: Sage, 2000.
Lillico, Ian. "The School Reforms Required to Engage Boys in Schooling." Boys Forward. http://www.boysforward.com/school%20reforms.htm (accessed January 16, 2008).
Lincoln, Yvonna S., and Gaile S. Cannella. "Dangerous Discourses: Methodological Conservatism and Governmental Regimes of Truth." *Qualitative Inquiry* 10, no. 1 (2004): 5–14.
Lingard, Bob. "Federalism in Schooling since the Karmel Report (1973), *Schools in Australia*: From Modernist Hope to Postmodernist Performativity." *Australian Educational Researcher* 27, no. 2 (2000): 25–61.
———. "Where to in Gender Policy in Education after Recuperative Masculinity Politics?" *International Journal of Inclusive Education* 7, no. 1 (2003): 33–56.

Lingard, Bob, and Peter Douglas. *Men Engaging Feminisms: Pro-Feminism, Backlashes, and Schooling.* Buckingham, England: Open University Press, 1999.

Lingard, Bob, Debra Hayes, Martin Mills, and Pam Christie. *Leading Learning: Making Hope Practical in Schools.* Maidenhead, England: Open University Press, 2003.

Lingard, Bob, Wayne Martino, Martin Mills, and Mark Bahr. *Addressing the Educational Needs of Boys.* Canberra, Australia: Department of Education, Science and Training, 2002.

Lingard, Bob, and Paige Porter, eds. *A National Approach to Schooling in Australia? Essays on the Development of National Policies in Schools Education.* Canberra: Australian College of Education, 1997.

Lopez, Nancy. *Hopeful Girls, Troubled Boys.* New York: Routledge, 2003.

Lopez-Claros, Augusto, and Saadia Zahidi. *Women's Empowerment: Measuring the Global Gender Gap.* Geneva, Switzerland: World Economic Forum, 2005.

Mac an Ghaill, Maírtín. "(In)Visibility: Sexuality, Race, and Masculinity in the School Context." In *Challenging Gay and Lesbian Inequalities in Education*, edited by Debbie Epstein, 152–76. Buckingham, England: Open University Press, 1994.

———. *The Making of Men: Masculinities, Sexualities, and Schooling.* Buckingham, England: Open University Press, 1994.

———. "'New Times' In an Old Country: Emerging Black Gay Identities and (Hetero) Sexual Discontents." In *Masculinities at School*, edited by Nancy Lesko, 163–85. Thousand Oaks, CA: Sage, 2000.

Macintyre, Stuart. *A Concise History of Australia.* Cambridge, England: Cambridge University Press, 1999.

Madden, Mary L. "Which Boys? Which Girls? Gender, Social Class, and the Politics of the 'Boy Crisis.'" Paper presented at the annual meeting of the American Educational Research Association, Chicago, April 9, 2007.

Magarey, Kirsty. *Sex Discrimination Amendment (Teaching Profession) Bill 2004.* Canberra, Australia: Department of Parliamentary Services, 2004.

Maguire, Meg, and Stephen J. Ball. "Researching Politics and the Politics of Research: Recent Qualitative Studies in the UK." *Qualitative Studies in Education* 7, no. 3 (1994): 269–85.

Maine Task Force on Gender Equity in Education. *Draft Report.* Augusta: Maine Department of Education, 2006.

———. *Final Report.* Augusta: Maine Department of Education, 2007.

Marceau, Jane. *Steering from a Distance: International Trends in the Financing and Governance of Higher Education.* Canberra: Australian Department of Employment, Education, and Training, 1993.

Marshall, Catherine. "Policy Discourse Analysis: Negotiating Gender Equity." *Journal of Education Policy* 15, no. 2 (2000): 125–56.

Martino, Wayne. "'Dickheads, Wuses, and Faggots': Addressing Issues of Masculinity and Homophobia in the Critical Literacy Classroom." In *Negotiating Critical Literacies in Classrooms*, edited by Barbara Comber and Anne Simpson, 171–87. Mahwah, NJ: Lawrence Ehrlbaum, 2001.

———. "It's Not the Way Guys Think!" In *Boys in Schools: Addressing the Real Issues—Behaviour, Values and Relationships*, edited by Rollo Browne and Richard Fletcher, 124–38. Sydney: Finch, 1995.

Martino, Wayne, and Deborah Berrill. "Boys, Schooling, and Masculinities: Interrogating the 'Right' Way to Educate Boys." *Educational Review* 55, no. 2 (2003): 99–117.

Martino, Wayne, and Michael Kehler. "Male Teachers and the 'Boy Problem': An Issue of Recuperative Masculinity Politics." *McGill Journal of Education* 41, no. 2 (2006): 113–31.

Martino, Wayne, Bob Lingard, and Martin Mills. "Issues in Boys' Education: A Question of Teacher Threshold Knowledges?" *Gender and Education* 16, no. 4 (2004): 435–54.

Martino, Wayne, and Maria Pallotta-Chiarolli. *So What's a Boy? Addressing Issues of Masculinity and Schooling*. Maidenhead, England: Open University Press, 2003.

Mayer, Richard E. "What Is the Place of Science in Educational Research?" *Educational Researcher* 29, no. 6 (2000): 38–39.

Maynard, Trisha. *Boys and Literacy: Exploring the Issues*. London: Routledge/Falmer, 2002.

McCracken, Nancy Mellin. "Surviving Shock and Awe: NCLB vs. Colleges of Education." *English Education* 36, no. 2 (2004): 104–18.

McKay, Jim, Geoffrey Lawrence, Toby Miller, and David Rowe. "Gender Equity, Hegemonic Masculinity and the Governmentalisation of Australian Amateur Sport." In *Culture in Australia: Policies, Publics and Programs*, edited by Tony Bennett and David Carter, 233–51. Cambridge, England: Cambridge University Press, 2001.

McLean, Christopher. "Boys and Education in Australia." In *Men's Ways of Being*, edited by Christopher McLean, Maggie Carey, and Cheryl White, 65–83. Boulder, CO: Westview Press, 1996.

McNeil, Linda. *Contradictions of School Reform: Educational Costs of Standardized Testing*. New York: RoutledgeFalmer, 2000.

Mead, Sara. "The Truth About Boys and Girls." Washington, DC: Education Sector, 2006.

Meinhof, Ulrike Hanna. "'The Most Important Event of My Life!' A Comparison of Male and Female Written Narratives." In *Language and Masculinity*, edited by Sally Johnson and Ulrike Hanna Meinhof, 208–28. Oxford, England: Blackwell, 1997.

Millard, Elaine. *Differently Literate: Boys, Girls and the Schooling of Literacy*. London: Falmer Press, 1997.

Mills, Martin. *Challenging Violence in Schools: An Issue of Masculinities*. Buckingham, England: Open University Press, 2001.

———. "Issues in Implementing Boys' Programme in Schools: Male Teachers and Empowerment." *Gender and Education* 12, no. 2 (2000): 221–38.

———. "Shaping the Boys' Agenda: The Backlash Blockbusters." *International Journal of Inclusive Education* 7, no. 1 (2003): 57–73.

Mills, Martin, Wayne Martino, and Bob Lingard. "Attracting, Recruiting and Retaining Male Teachers: Policy Issues in the Male Teacher Debate." *British Journal of Sociology of Education* 25, no. 3 (2004): 355–69.

———. "Getting Boys' Education 'Right': The Australian Government's Parliamentary Inquiry Report as an Exemplary Instance of Recuperative Masculinity Politics." *British Journal of Sociology of Education* 28, no. 1 (2007): 5–21.

Minister for Education Science and Training. *Boys' Education: Building on Successful Practice.* Canberra, Australia: Department of Education, Science and Training, 2003.

Ministerial Council on Education, Employment, Training and Youth Affairs Gender Equity Taskforce. *Gender Equity: A Framework for Australian Schools [the Gender Equity Framework].* Canberra, Australia: MCEETYA, 1997.

Ministerial Council on Education, Employment, Training and Youth Affairs. *The Adelaide Declaration on the Goals for Schooling in the Twenty-First Century.* Carlton, Australia: MCEETYA, 1999

Moir, Anne, and David Jessel. *Brain Sex: The Real Difference between Men and Women.* New York: Carol Publishing Group, 1991.

Moore, Robert L., and Douglas Gillette. *King, Warrior, Magician, Lover: Rediscovering the Archetypes of the Mature Masculine.* San Francisco: HarperSanFrancisco, 1990.

Morrison, Toni. *Playing in the Dark: Whiteness and the Literary Imagination.* New York: Vintage, 1992.

Munns, Geoff, Leonie Arthur, Toni Downes, Robyn Gregson, Anne Power, Wayne Sawyer, Michael Singh, Judith Thistleton-Martin, and Frances Steele. *Motivation and Engagement of Boys: Evidence-Based Teaching Practices.* Canberra, Australia: Department of Education, Science and Training, 2006.

Murtadha-Watts, Khaula. "Theorizing Urban Black Masculinity Construction in an African-Centered School." In *Masculinities at School,* edited by Nancy Lesko, 49–71. Thousand Oaks, CA: Sage, 2000.

National Center for Educational Statistics. *Condition of Education 2004.* Washington, DC: U.S. Department of Education, 2004.

———. *NAEP 2004 Trends in Academic Progress: Three Decades of Student Performance in Reading and Mathematics.* Washington, DC: U.S. Department of Education, 2005.

———. *The Nation's Report Card: Reading 2007.* Washington, DC: U.S. Department of Education, 2007.

———. *The Nation's Report Card: Writing 2007.* Washington, DC: U.S. Department of Education, 2008.

National Commission on Excellence in Education. *A Nation at Risk: The Imperative for Educational Reform.* Washington, DC: National Commission on Excellence in Education, 1983.

National Inquiry into the Teaching of Literacy. *Teaching Reading: Report and Recommendations.* Canberra, Australia: Department of Education, Science and Training, 2005.

National Office of Overseas Skills Recognition. *Country Education Profiles: Australia.* 3rd ed. Canberra, Australia: Department of Education, Training and Youth Affairs, 2000.

NEA Research. *Status of the American Public School Teacher 2000–2001.* Washington, DC: National Education Association, 2003.

New South Wales Government Advisory Committee on Education Training and Tourism. *Challenges and Opportunities: A Discussion Paper.* Sydney: New South Wales Ministry of Education & Youth Affairs, 1994.

Newkirk, Thomas. "'Brain Research'—a Call for Skepticism." *Education Week,* October 12, 2005, 29.

———. *Misreading Masculinity: Boys, Literacy, and Popular Culture*. Portsmouth, NH: Heinemann, 2002.

———. "Misreading Masculinity: Speculations on the Great Gender Gap in Writing." *Language Arts* 77 (2000): 294–300.

Newman, Katherine S. *Rampage: The Social Roots of School Shootings*. New York: Basic Books, 2004.

Noble, Colin, and Wendy Bradford. *Getting It Right for Boys and Girls*. London: Routledge, 2000.

Orenstein, Peggy. *School Girls: Young Women, Self-Esteem, and the Confidence Gap*. New York: Anchor Books, 1994.

Organisation for Economic Cooperation and Development. *Education at a Glance: OECD Indicators 2003*. Paris: OECD, 2003.

Oswald, Donald P., Al M. Best, Martha J. Coutinho, and Heather A. L. Nagle. "Trends in the Special Education Identification Rates of Boys and Girls: A Call for Research and Change." *Exceptionalities* 11 (2003): 223–37.

Pidgeon, Sue. "Learning Reading and Learning Gender." In *Reading the Difference: Gender and Reading in Elementary Classrooms*, edited by Myra Barrs and Sue Pidgeon, 20–37. York, ME: Stenhouse, 1994.

Pipher, Mary. *Reviving Ophelia: Saving the Selves of Adolescent Girls*. New York: Grosset/Putnam, 1994.

Pirie, Bruce. *Teenage Boys and High School English*. Portsmouth, NH: Heinemann, 2002.

Pollack, William. *Real Boys: Rescuing Our Sons from the Myths of Boyhood*. New York: Random House, 1998.

Pressley, Michael. *Reading Instruction That Works: The Case for Balanced Teaching*. New York: Guilford Press, 1998.

Prunty, John J. "Signposts for a Critical Educational Policy Analysis." *Australian Journal of Education* 29, no. 2 (1985): 133–40.

Rigby, Ken. *A Meta-Evaluation of Methods and Approaches to Reducing Bullying in Pre-Schools and in Early Primary School in Australia*. Canberra, Australia: Commonwealth Attorney General's Department, 2002.

Ross, Tara. "Nats Call for Inquiry into Boys' Education." Fairfax New Zealand Ltd. http://www.stuff.co.nz/stuff/0,2106,2668521a7694,00.html (accessed October 27, 2003).

Rowan, Leonie, Michele Knobel, Chris Bigum, and Colin Lankshear. *Boys, Literacies and Schooling: The Dangerous Territories of Gender-Based Literacy Reform*. Buckingham, England: Open University Press, 2002.

Rowden-Racette, Kellie. "Endangered Species." *Teacher Magazine,* November 1, 2005.

Rowe, David, and Jim McKay. "A Man's Game: Sport and Masculinities." In *Male Trouble: Looking at Australian Masculinities*, edited by Stephen Tomsen and Mike Donaldson, 200–16. Melbourne, Australia: Pluto Press, 2003.

Sabatier, Paul A. *Theories of the Policy Process*. 2nd ed. Cambridge, MA: Westview Press, 2007.

Sadker, David. "An Educator's Primer on the Gender War." *Phi Delta Kappan*, November 2002.

Sadker, Myra, and David Sadker. *Failing at Fairness: How Our Schools Cheat Girls*. New York: Touchstone, 1994.

Salisbury, Jane, and Sheila Riddell. *Gender, Policy & Educational Change: Shifting Agendas in the UK and Europe.* London: Routledge, 2000.

Salisbury, Jonathan, and David Jackson. *Challenging Macho Values: Practical Ways of Working with Adolescent Boys.* London: Falmer Press, 1996.

Sargent, Paul. *Real Men or Real Teachers? Contradictions in the Lives of Men Elementary School Teachers.* Harriman, TN: Men's Studies Press, 2001.

Saunders, Kay, and Geoffrey Bolton. "Girdled for War: Women's Mobilisations in World War Two." In *Gender Relations in Australia: Domination and Negotiation*, edited by Kay Saunders and Raymond Evans, 376–97. Sydney: Harcourt Brace, 1992.

Saunders, Kay, and Raymond Evans, eds. *Gender Relations in Australia: Domination and Negotiation.* Sydney: Harcourt Brace, 1992.

Sax, Leonard. *Why Gender Matters: What Parents and Teachers Need to Know about the Emerging Science of Sex Differences.* New York: Random House, 2006.

Sewell, Tony. *Black Masculinities and Schooling: How Black Boys Survive Modern Schooling.* Staffordshire, England: Trentham Books, 1997.

Sexton, Patricia. *The Feminized Male: Classrooms, White Collars and the Decline of Manliness.* New York: Random House, 1969.

Sheridan, Dorothy, Brian Street, and David Bloome. *Writing Ourselves: Mass-Observation and Literacy Practices.* Cresskill, NJ: Hampton Press, 2000.

Simmons, Rachel. *Odd Girl Out: The Hidden Culture of Aggression in Girls.* New York: Harcourt, 2002.

Simms, Marian. "Two Steps Forward, One Step Back: Women and the Australian Party System." In *Gender and Party Politics*, edited by Joni Lovenduski and Pippa Norris, 16–34. London: Sage, 1993.

Singleton, Gwynneth, Don Aitkin, Brian Jinks, and John Warhurst. *Australian Political Institutions.* 5th ed. South Melbourne, Australia: Longman, 1996.

Skelton, Christine. *Schooling the Boys: Masculinities and Primary Education.* Buckingham, England: Open University Press, 2001.

Slade, Malcolm. *Listening to the Boys: Issues and Problems Influencing School Achievement and Retention.* Adelaide, Australia: Shannon Research Press, 2002.

Smith, Mary Lee. *Political Spectacle and the Fate of American Schools.* New York: RoutledgeFalmer, 2004.

Smith, Michael W., and Jeffrey D. Wilhelm. *Reading Don't Fix No Chevys: Literacy in the Lives of Young Men.* Portsmouth, NH: Heinemann, 2002.

Smith, Rosa A. "Black Boys." *Education Week*, October 30, 2002.

Sommers, Christina Hoff. *The War against Boys: How Misguided Feminism Is Harming Our Young Men.* New York: Simon & Schuster, 2000.

———. *Who Stole Feminism?: How Women Have Betrayed Women.* New York: Simon & Schuster, 1994.

Taylor, Sandra, Fazal Rizvi, Bob Lingard, and Miriam Henry. *Educational Policy and the Politics of Change.* London: Routledge, 1997.

Thompson, Michael. *Speaking of Boys: Answers to the Most-Asked Questions About Raising Sons.* New York: Ballantine Books, 2000.

Thomson, Pat. *Schooling the Rustbelt Kids: Making the Difference in Changing Times.* Crows Nest, Australia: Allen & Unwin, 2002.

Thorne, Barrie. *Gender Play: Girls and Boys in School*. New Brunswick, NJ: Rutgers University Press, 1993.
Titus, Jordan J. "Boy Trouble: Rhetorical Framing of Boys' Underachievement." *Discourse: Studies in the Cultural Politics of Education* 25, no. 2 (2004): 145–69.
Trent, Faith, and Malcolm Slade. *Declining Rates of Achievement and Retention: The Perceptions of Adolescent Males*. Canberra, Australia: Department of Education, Training and Youth Affairs, 2001.
Tyre, Peg. "The Trouble with Boys." *Newsweek*, January 30, 2006.
U.S. Department of Education. *Single-Sex Versus Coeducational Schooling: A Systematic Review*. Washington, DC: U.S. Department of Education, 2005.
United Nations Children's Fund (UNICEF). *The State of the World's Children 2004*. New York: United Nations Children's Fund, 2004.
van Acker, Elizabeth. *Different Voices: Gender and Politics in Australia*. South Yarra, Victoria, Australia: Macmillan Education, 1999.
van Dijk, Teun A. "Discourse and the Denial of Racism." In *The Discourse Reader*, edited by Adam Jaworski and Nikolas Coupland, 541–58. London: Routledge, 1999.
Von Drehle, David. "The Boys Are All Right." *Time*, August 6, 2007.
Walker, Vanessa Siddle. "After Methods, Then What? A Researcher's Response to the Report of the National Research Council." *Teachers College Record* 107, no. 1 (2005): 30–37.
Ward, Russel. *The Australian Legend*. 2nd ed. Melbourne, Australia: Oxford University Press, 1958.
Weaver-Hightower, Marcus B. "The 'Boy Turn' in Research on Gender and Education." *Review of Educational Research* 73, no. 4 (2003): 471–98.
———. "Crossing the Divide: Bridging the Disjunctures between Theoretically Oriented and Practice-Oriented Literature About Masculinity and Boys at School." *Gender and Education* 15, no. 4 (2003): 407–23.
———. "Dare the School Build a New Education for Boys?" *Teachers College Record* (2005), http://www.tcrecord.org/content.asp?contentid=11743 (accessed January 16, 2008).
———. "An Ecology Metaphor for Educational Policy Analysis: A Call to Complexity." *Educational Researcher* 37, no. 3 (2008): 153–67.
———. "Every Good Boy Does Fine: Policy Ecology, Masculinity Politics, and the Development and Implementation of Australian Policy on the Education of Boys, 2000–2005." PhD diss., University of Wisconsin-Madison, 2006.
———. "Inventing the 'All-American' Boy: A Case Study in the Capture of Boys' Education Issues by Conservative Groups." *Men and Masculinities* 10, no. 3 (2008): 267–95.
Weis, Lois. *Class Reunion*. New York: Routledge, 2004.
———. *Working Class without Work*. New York: Routledge, 1990.
Williams, Raymond. *Key Words*. New York: Oxford University Press, 1985.
———. *Marxism and Literature*. Oxford, England: Oxford University Press, 1977.
Willis, Paul. *Learning to Labor*. New York: Columbia University Press, 1977.

Wiseman, Rosalind. *Queen Bees and Wannabes: Helping Your Daughter Survive Cliques, Gossip, Boyfriends, and Other Realities of Adolescence*. New York: Crown Publishers, 2002.

Women on Words and Images. *Dick and Jane as Victims: Sex Stereotyping in Children's Readers*. Princeton, NJ: Women on Words and Images, 1972.

Woodside-Jiron, Haley. "Critical Policy Analysis: Researching the Roles of Cultural Models, Power, and Expertise in Reading Policy." *Reading Research Quarterly* 38, no. 4 (2003): 530–36.

Yates, Lynn. "The 'Facts of the Case': Gender Equity for Boys as a Public Policy Issue." In *Masculinities at School*, edited by Nancy Lesko, 305–22. Thousand Oaks, CA: Sage, 2000.

Young, Josephine Peyton. "Boy Talk: Critical Literacy and Masculinities." *Reading Research Quarterly* 35, no. 3 (2000): 312–37.

Index

Page numbers in **bold** print refer to figures or tables.

3R program, 84

Aboriginal Australians, *see* Indigenous Australians
academics and researchers, 24, 65, **71**, 77, 85, 89, 95, 100, 101, 105, 109, 114, 129, 136, 139, *see also individual names*
accommodation, 88–9
accountability, 11, 18, 22, 115–16, 121, 177, 187, 190, 224 (note 50)
action research, 129, 139, 141, 149, 150–2, 153–4, 155
Adelaide Declaration, the, 114
affinity groups, 140–1, 142, 190
African American boys, xii, 19, 24, 180, 184, 185, 188–9, 194
agreement score, 70, 72, **73–6**
alliance building, 4, 202, 203
Alloway, Nola (academic, director of BELS phase one), 115, 127, 135
alternative schools, 6, **8**, 68, 84, 106
American Association of University Women (AAUW), 181–2, 187, 191, 201
Anglin, Doug, 191–2
anti-bullying, 92, 106, 201
antifeminism, *see* backlashes against feminism
antiracism, 193
antisexism, 92, 121, 193, 204
antiviolence, 92, 193, 204
Apple, Michael (academic), xi–xiv, 52, 87, 96, 102, 106, 197, 201, 203
arguments
 made in *BGIR*, 5–6, 68–9, 70, **71**
 made to the Committee, 68–72, **73–6**
 reduced list used in submission analysis, 70–1, **71**
Asian peoples and Asian-Australians, 35

assessment, *see* tests and testing
athletics, *see* sport
attention deficit disorder, 184
auditory processing, 6, **8**, 83–4
 compare hearing (auditory)
Australia
 currency, xix
 gender relations, 36–8
 government and policymaking structure, 43–5, *see also* Commonwealth government
 map, **33**
 policy ecology, educational, 25–6, 55, 80, 87–8
 population, 25, **33**, 34, 35, **132**, 187
 race relations, 1, 34–5
 rurality, 34, 78
 size (land mass), 34
 states, *see individual states*
 why study, 3–4
Australian Association of Social Workers, **73**, 78
Australian Bureau of Statistics, 6, **8**, **33**
Australian Capital Territory, **33**, **132**
Australian Council of Deans of Education, 125
Australian Council of State School Organisations, **73**, 78
Australian Education Union, **73**, 78–9, 85, 95, *see also* unions, teachers
Australian Hearing, **73**, 77, 94
Australian Labor Party (ALP), 12, 26, 32, 45, 53, 65, 67, 89, 118, 126, 129
 move to the political right, 53, 67, 89–90
Australian Schools Commission, 45–6
awards, 163, 168

backlashes against feminism, 2, 22–3, 26, 31, 37, 49, 67, 79, 85, 96, 98, 101, 120, 138, 142, 172, 183, 190, 191, 197, 205, *see also* recuperative masculinity politics
Ball, Stephen, 9, 10, 59, 103, 171
Barresi, Phillip (Committee member), 73, 86, 92–3
Bartlett, Kerry (Committee chair), 62, 63, 67, **73**, 95–6, **207**
behavior, 6, **8**, 19, 39, 60, 92, 99, 109, 147, 168, 184, *see also* discipline
Biddulph, Steve (author and consultant), 21, 101, 169, 170
Bishop, Julie (former minister of education), 200
Black boys, *see* African American boys; indigenous Australians, boys
Bligh, William (naval captain and Governor of Australia), 31
books on boys' education
popular-rhetorical, 21, 92, 101, 169, 171, 182–3
practice-oriented, 92, 127, 179, 182, 203, 204
theoretically oriented, 20, 21, 169–70, 182–3, 203
boy band, 104
boy crisis, *see* panics over boys
"boy friendly" practices, 50, **71**, 110, 205
Boys: Getting It Right (BGIR), 3, 5–6, 7, 10, 11, 12, 14, 17, 25, 30, 31, 34, 35–6, 38, 39, 42, 43, 44, **46**, 49, 50–2, 55, 60, 62, 64, 68, 70–2, 77, 78–84, 87, 89–90, 94–5, 96–101, 102, 103, 104, 105, 110–11, 113, 114–15, 121, 123–4, 125, 128, 138, 140, 145, 161, 162, 167–8, 169–71, 173, 179, **207**
arguments made in, 5–6, 68–9, 70, **71**
assertions of legitimacy, 12, 52, 59, 62, 64, 94–5, 101, 105, 138
conflicts of interest, 101
as conservative, 98, 100, 102, 105
cover, 12, **13**
citations, explicit, 72, **73–6**
as definer of acceptable boys, 102
elements of bad sense, 98–101
elements of good sense, 96–8
impact on girls' education, 98–9
influences, most significant, 77–86
as masculinity politics, 14
neglect of masculinity, homophobia, sexism, violence and racism, 99
as "policy" rather than "report," 10–14, 47, 114
recommendations made in, 5–6, **7–8**, 70
as unanimous, 66, 67, 89, 97
use of by educators and schools, 146, 162, 167–8, 169–71
as validation of consultants, 105
writing of, 62–4
boys as gendered, 97, 198
boys as neglected, 104–5, 109
Boy Scouts, 181
boys' education
centrality to conservative modernization, 197
coercion to participate in, 156, 174
curiosity about, 171–2
divisiveness and tensions created by, 163, 166, 175, 176, 193
experiences of failure with, 174–5
as inherently political, 66–7, 70, 129
as not inherently conservative, 24, 53, 194
as opportunity for career advancement, 173
teachers' reasons for not participating, 174–6
teachers' reasons for participation, 171–4
as white, middle-class concern, 24, 53, 85, 101, 185
Boys' Education: Building on Successful Practice, *see* Government response
Boys' Education Lighthouse Schools (BELS) program, 34, 92, 101, 103, 106, 112, 127–35, 136, **137**, 138, 139, 140, 141, 142–154, 198, **207–8**
allotment by state and sector, 129, **132**, 133
as politically shaped, 128–34, 221 (note 57)
reports, 127–8, 154, **208**
see also Riverside Schools Cooperative
Boys in Focus, **73**, 79, 85
Boys to Fine Men (B2FM) conferences (Newcastle 2003, Melbourne 2005), 82, 83, 104–12, 136, **137**, 140, 159, 162, 173, **208–9**
attendance, 106
bodily pleasures, 104, 106, 173
celebrating successes, 106, 108, 111–12
commercialization, 107, *see also* products and services
as conservative, 107, 109–12

INDEX / 245

costs, 107–8
focus on practice and practitioners, 106, 111
government and corporate funding, 108
indigenous activism and participation, 108–9
as multimedia event, 104, 106
(pro)feminist participation and impact, 106, 108–9, 112
teasers for professional development workshops, 107
boys' underachievement, 3, 5, 97
in English, 18
boy turn, 17–26, 55, 91, 128, 180, 182, 197, 203
brain-based explanations of gender differences, 18, 101, 152, 177, 182
briefing notes, 62, 65
bullying, 19, 92, 93, 97, 99, 106, 198, 202, 212 (note 11)
Bush, George W. (U.S. president) and his administration, 191, 192, 193
business model, influence on schooling, 148, 156

Canada, 17, 56, 142–3, 179, 187
Canberra Grammar School, **73**, 79–80
Catholic Education Office, Archdiocese of Sydney, 125, **207**, **208**
Challenges and Opportunities: A Discussion Paper, see O'Doherty Report
character education, 182
cherry picking, *see under* policy
choice, school, 22, 177, 187, 193, *see also* vouchers
Christian conservatism, 142, 177, 190–1
Clarke, Edward, 181
class, social, xii, xiii, 18, 20, 24, 47, 53, 55, 64, 65, 85, 95, 101, 147, 148, 151, 152, 157, 183–4, 185, 190, 194, 197, 202, 205, *see also* socioeconomic status
class size, 6, 7, **8**, **71**, 79, 84, 97
Clinton, Hillary, 200
Coalition Government, *see* Liberal-National Coalition
coeducation, 160–1, 164, 166, 176, 180–1
colonial history and impact, 24, 30–1, 55–6
Committee on Education and Training, House of Representatives Standing, xix, 5, 6, 10, 14, 36, 39–40, 42, 50–1,
54, 59, 60–70, 72, **73–6**, 77–87, 90–1, 94–5, 97–8, 99–100, 102, 105, 113, 114, 123, 128, **207**
arguments made to the, 68–72, **73–6**
attendance at public hearings, 65–6
beliefs of members, a priori, 63, 86, 95
carryover of members, 65
composition by political party, 64–6, *see also* party politics
deliberations within, 63–4
knowledge of issues by members, 62
"mishearings" by, 90–1
as unqualified to validate research, 94–5, 105
women members' participation in, 63–4
see also Inquiry into the Education of Boys
Commonwealth government
bureaucracy, 37, 45
committee system, 45, 60–8
funding for boys' education, 136, **137**, 140
role in education, 6, 33, 34, 54
tensions with states and territories, 6, 43, 44, 100
see also Department of Education, Science and Training; Department of Education, Training and Youth Affairs; gender and education policies, Australian; Ministerial Council for Education, Employment, Training and Youth Affairs
community service centers, 78
comparability among the states, **8**, 44, 54, 114, 115
competition between schools, 22, 147, 148, 156, 168, 177, *see also under* private schools; public schools
compulsory heterosexuality, 41–2
computers, *see* information and communication technologies
conservative groups and movements, 61, 112, 116, 129, 140, 142, 190, 197, 202
conservative ideology and politics, 24, 52, 53, 85–6, 87, 101, 106, 108, 109, 116, 123, 169, 177, 182–3, 190
consultants, 101, 105, 106, 107, 121, 129, 136, 138, 142, 169, 173, 204
co-opting of progressive discourses by conservatives, 89

counseling, 158, 189, 204
counterintuitive big ask, 92–3
crisis of masculinity, *see under* masculinities
critical literacy, 20, 92, 98
crocodile hunter image, 30, 38
cross-tabulation of analysis methods, 73–6, 77
culture, school, 71, 118, 146, 159–61, 155, 156, 166–7, 168, 175, 176
curriculum, 5–6, 18, 22, 47, 49–50, 52, 54, 62, 66, 71, 84, 86, 99, 100, 119, 120, 138, 152, 171, 181, 182, 204
Cuttance, Peter (director of BELS phase two), 127, 221 (note 57)
cycle of legitimacy, 94–5, 101, 105, *see also* legitimacy

dance, 108, 177, 204
de facto policy, 11, 12, 92, 121, 158, 170, 183, 188
deficit model, 90–1, 122, 155
Delfos, Martine (academic), 109
Department of Education, Science and Training (DEST), 10, 44, 54, 74, 80–1, 85, 103, 108, 113, 114, 115, 117, 118, 123, 124, 127–8, 136, 139, 200, 208–9
Department of Education, Training and Youth Affairs (DETYA), *see* Department of Education, Science and Training
depression, *see* mental health
Detroit all-male schools, 186, 189
disabilities, 134, 185, *see also* attention deficit disorder; mental health; special education
disaffection from schooling, boys', 84
discipline, 8, 66, 82, 94, 99, 111, 162, 184
discourse flows, *see* globalization of educational discourses
division of labor, gendered, 37, 121, 155, 167, 175
Dobson, James, 182, 188
DREAMS program, 189
drug and alcohol abuse, 2, 19, 51, 53, 184
drunk driving, 19, 97, 99

economic indicators and concerns for boys, 19, 23–4, 53, 184
economy of boys' education, 107–8, 117, 128
education as moral savior, 31

education departments, state and territory, 42, 71, 79, 85, 100, 199
Education Queensland (EQ), 74, 81–2
Education Sciences Reform Act of 2002 (United States), 100
electioneering, 126, 133–4, 138
"endgame" in girls policy, 50, 51, 123, 198
England, 17, 22, 25, 30–2, 35, 41, 43, 56, 127, 179, 187, 190, 194
"equality," xi, 31, 52, 102
evidence-based research, 112
examinations, *see* tests and testing
exhibits submitted to the Inquiry, *see under* Inquiry into the Education of Boys
exporting blame, 6, 43, 100, 138–9

Faludi, Susan, 23–4
fathers, 6, 8, 71, 82, 106, 109, 137, 163, 188, 189, 209
federal government, *see* Commonwealth government
federalism, 4, 33, 43–4, 47, 100, 185–6
female teachers, *see under* teachers
feminists, 2, 20, 21, 22, 37, 38, 46, 47, 50, 52, 101, 112, 175, 182, 191, 213 (note 26)
alliances with men and homosexual groups, 89
in the bureaucracy, *see* femocrats
responses to boy panics, 2, 20
role in boy turn, 21
strategic mistakes, 22
successes, 37, 52
see also (pro)feminists
feminization
of boys by female teachers, 181, 199
of reading, 100
of schools, 18, 47, 50
of teaching, 100, 185
of workforce and workplace, 23, 24
femocrats, 37, 38, 45, 47, 49, 67, 112, 138, 199
Ferguson, Ann Arnett, 185
Fletcher, Richard, 32, 41, 74, 79, 82, 85, 93, 101, 105, 110, 115, 117, 145, 169
funding for boys' education, 17, 67, 98–9, 101, 127, 128, 146, 148, 154, 156, 172, 173–4, 176, 200, 204
funding for girls' education, 197–8, 201, 204

GaiSheridan International (GSI), 115–23, **208**
gangs, 79, 192

"gender," use of the term instead of "girls," 48, 50, 51, 55, 56
gender and education policies
 Australian, 37, 45–52, **46**, 102
 continuing viability of for girls, 199
 U. S., 180–3
 see also individual titles
gender as relational, 49, 50, 51, 193
gender as social construction, 48, 90, 91, 98, 117–18, 119–20, 122, 153, 185
gender as women's issue, 187
Gender Equity Framework (*Gender Equity: A Framework for Australian Schools*), 7, 12, **46**, 50, 55, **71**, 98, 99, 100, 113–23, 138, 139, 176
 recasting, 12, 26, 50, 51, 82, 92, 99, 111, 113–23, **137**, 138, **208**
 see also GaiSheridan International
gender equity indicators, *see* indicators of advantage or disadvantage
Gender Equity in Education Act (U.S.), 182
gender neutrality, 175
gender relations, *see under* Australia
gender segregation, 46, 166–7
Germany, 17
Gillard, Julia (former Committee member, Minister of Education), 54, 61, 89, 199–200, **207**, 216 (note 81)
girls
 culling of research on, 139
 education policy for, *see* "endgame" in girls policy; gender and education policies
 focus on in the U.S., 181–2
 programs for, allegiance to, 151, 175, 201
 progressive policy and programs for, 199, 201
 social and economic outcomes, 153, 193, 205
Girls, School and Society (report), 46, **46**, 47, 48
Girls and Tomorrow (report), **46**, 47
globalization of educational discourses, 4, 21, 25–6, 55, 69, 168
Government response, 6–8, 10, 12, 44, **46**, 113, 124, **208**
Gretna Green High School, *see* Riverside Schools Cooperative
"Guiding Principles for Boys' Education," 127–8
Gurian, Michael (author and consultant), 142, 177, 182

Hall, G. Stanley, 181, 199
Hansard transcripts, xix, 61, 212 (note 20)
Hartman, Deborah, 110, 112, 115, 117
hearing (auditory), 6, **8**, 77, 94
hearing and vision screening, 7, **8**, 77, 78, 98
hearings, public, *see under* Inquiry into the Education of Boys
hegemonic masculinity, *see under* masculinities
heterosexuality, compulsory, 41–2
higher education, 5, 18–19, 20, 54, 148, 184
Higher Education Contribution Scheme (HECS), **8**, 123, 124
Hill, Peter (academic), 74, 82–3, 85
Hispanic boys, 184
homophobia, 19, 41, 42, 86, 91, 99, 109, 118, 121, 123, 193, 202
homosexuality, 41, 52, 188
hope, 197–202
Howard, John (former prime minister), 65, 126
How Schools Shortchange Girls, 181–2, *see also* American Association of University Women
Human Rights and Equal Opportunity Commission (HREOC), 125, **207**

Iceland, 3, 17, 20, 143, 179
In a Different Voice (Carol Gilligan), 25–6
indicators of advantage or disadvantage, 2, 5, 22, 50, 51, 52, 60, **71**, 94, 113, 114, 122, 163, 168, 184, 193
Indigenous Australians, 1–2, 5, 30–1, 34, 46, 52, 81, 83, 98, 99, 108–9, 134, 135, 147
 boys, 99, 108–9, 134, 135, 198
 Boys to Fine Men Forum, 108–9
 land rights, 34
influence, direction of, 25–6, 68
information and communication technologies (ICTs), 2, **71**, 93, 128, 135, 141, 147, 193, 199
Inquiry into the Education of Boys
 allocation of legitimacy, 62
 exhibits submitted to, 61–2, 217 (note 3)
 hearings, public, 61, 62, 65–6, 68, 69–70, **73–6**, 114, **207**
 process, 60–4, **207–9**
 submissions, written, 61, 62, 70–2, **73–6**, 114

Inquiry into the Education of Boys—
  continued
  terms of reference, 60–1, 69, 77, 91, **207**
  witness time allotment, 72, **73–6**, 77,
    217 (note 11)
  writing of the report, 62–4
  see also Boys: Getting It Right;
    Committee on Education and
    Training, House of Representatives
    Standing
international attention to boys, *see
  individual countries*
Invergowrie Foundation, 201

Japan, 17
Johnson, Richard, xiii
judicial precedence as policy, 186, 195

Karmel Report, 46, **46**
Kemp, David (former minister of
  education), 54, 60, 61, **207**

Labor party, *see* Australian Labor Party
language use, shifts in, 118, 119–20, *see
  also* "gender," use of the term instead
  of "girls"
larrikins and larrikinism, 39
Latham, Mark (Opposition leader,
  2004–05), 35, 53, 126
learning styles, 7, **71**, 78, 92, 128,
  149, 182
legitimacy, 10, 12, 52, 59, 62, 64, 89,
  94–5, 105, 106, 108, 112, 136, 140,
  see also cycle of legitimacy
Liberal-National Coalition (majority party,
  1996–2007), 65, 126, 129, 133, 200,
  207, **209**
Liberal Party, *see* Liberal-National
  Coalition
Lillico, Ian (consultant), 75, 77, 169, 170
Lingard, Bob (academic), 11, 23, 49, 50,
  112, 115, 121, 141, 202
Listening to Girls (report), **46**, 48
literacy, 2, 5, 6, **7–8**, 18, 25, 53, 60, 80, 82,
  83, 87, 94, 106, 109, 122, 128, 134,
  135, **137**, 149, 148, 150, 151, 152–4,
  155, 183, 193, *see also* critical literacy
local responsibility for gender equity, 12,
  51, 116, 121, 138–9, 169, 177, 178,
  180, 188–9, 220 (note 32)

Maine (U.S.), policymaking for boys in,
  56, 186, 199

Making Up Lost Time in Literacy
  (MULTILIT), 105, 134
male teachers and role models, 2, 6, 20, 49,
  71, 79, 81, 82, 86, 93, 100, 123–6,
  128, 135, 143, 149, 153, 155–6,
  179–80, 184–5, **208**, *see also*
  scholarships for male teachers
marketing schools with boys' education,
  156, 168
Martino, Wayne, 39, 40, 115, 151, 203
masculinities, xii, 3, 20, 30, 38, 80,
  81, 187
  African American, 185
  Australian, 29, 30, 38–43
  Christian, 187, 188, 190
  colonial, 55–6
  competing ideologies of, 188
  creation of through policy, 30
  crisis of, 23–4, 26, 53, 126, 184,
    193, 199
  emotionality and, 96
  globalized, 18, 55–6
  harmful effects of certain, 97, 99, 198
  hegemonic, 30, 41, 81, 99, 119, 214
    (note 1)
  homosexual, 41–2
  materiality of, 30, 53
  media images of, 5, 30, 61, **71**, 79, 97, 98
  nationalized, 38
  need to change, 80, 81, 109, 198
  politics, 4, 14, 50, 202, *see also*
    recuperative masculinity politics
  race and, 19, 20
  social class and, 20
  therapy model, 188, 212 (note 15 under
    chapter 1)
  typologies of, 20, 30, 151
materials for boys' education, *see* products
  and services
mateship, 39–40
math, 2, 60, 121, 182, *see also* numeracy
media attention to boys' education, 21, 35,
  104, 169, 170, 184, 191
media literacy, critical, 98
Meeting the Challenge (BELS phase one
  report), 127–8, 154, **208**
men's movements, 21, 140, 187–8
  divorce and custody, 187
  lack of institutional infrastructure, 187
  Muscular Christianity, 190
  mythopoetic, 22, 187, 213 (note 29)
  see also Promise Keepers
mental health, 19, 82, 182

military and militarism, 21, 40–1, 181
millennial kids, 164–5, 166
Mills, Martin, 115–17, 150–1, 154, 155, 170, 204
Ministerial Council for Education, Employment, Training and Youth Affairs (MCEETYA), **8**, 34, 44, 48, 113, 114, 116, 118, 123, 127, **208**
money, *see* funding for boys' education
motivation, boys', 7, 109, **137**, 149, 152

National Action Plan for the Education of Girls, The, **46**, 48
National Inquiry into the Teaching of Literacy, 7, 12, 83
National Party, *see* Liberal-National Coalition
National Policy for the Education of Girls, 37, **46**, 47–8, 50, 55
Nation at Risk, A, 11–12, 181, 187
negative portrayal of or approach to boys, 5, **71**, 79, 99, 110–11, *see also* "positive" treatment of boys
Nelson, Dr. Brendan (former Committee chair, former minister of education, Leader of the Opposition), 6, 54, 65, **75**, 90, 115–16, 124–5, 200, **207–9**
neoconservative views of boys' education, 50, 182, 183
neoliberal educational reforms, 22, 168, 177, *see also* choice, school; vouchers
New Right, 22, *see also* conservative ideology and politics; neoconservative views of boys' education; neoliberal educational reforms
New South Wales, 11, 33, 48, 61, 81, **130–1**, **132**
New South Wales parliamentary inquiry into boys' education, *see O'Doherty Report*
New Zealand, 25, 179, 187
No Child Left Behind Act (U. S.), 100, 139, 142, 177, 187, 190, 193–4, 224 (note 50)
Northern Territory, 33, 35, 83, **132**
Northern Territory Department of Employment, Education and Training, **75**, 83
numeracy, 6, **7–8**, *see also* math

O'Doherty Report, **46**, 48–50, 61
outdoor education, 159, 167, 204

Paige, Rod (former U.S. secretary of education), 192
panics over boys, 3, 43, 53, 55, 80, 175, 179, 181, 192, 199
causes and origins, 21–5, 80
as "common sense," 88, 90, 177–8, 197
cyclical nature of, 198–9
parents and parenting, 7, 24, 53, 91, 93, 94, 95–6, 119, 122, 140, 148, 153, 156, 161, 168, 169, 172, 183, 185, 192, 193, 203, 205, **208**
cultural, racial and religious basis of, 95–6
as epistemological filter, 69, 95–6
single, 5, 20, 78, 98
targeting for progressive ends, 203, 205
partisanship, *see* party politics
party politics, 65, 66–8, 89
Paul Robeson Institute (Cambridge, Massachusetts), 189
pedagogy, 5–6, **8**, 47, 62, 66, 67, **71**, 80, 84, 92, 97, 98, 100, 106, 109, 119, 122, 138, 141, 151, 155, 182
pendulum metaphor, 175
phonics, 2, 6, 7, **8**, 11, **71**, 82–3, 98, 100, 105, 136
physical education, 31, 40, 181, 204
place-sensitive analysis and strategy, 202, 203, 206, *see also* situated strategizing
policy, 8–14, 101–2, 103
administrators as gatekeepers of, 168–9
analysis, 9–10, 56, 185, 206, 217 (note 10)
borrowing from other countries, 32
"cherry picking" from, 170–1, 177, 188, 201
de facto, 11, 12, 92, 121, 158, 170, 183, 188
ecologies, xiii, 9–10, 25, 26, 27, 52–5, 69, 80, 87–8, 180, 185, 195, 203, 206
education, *see* gender and education policies
independence from, 146, 158, 166, 173
levers, 9, 103, 190
rational model (stages heuristic), 8–9
regulatory, 177
symbolic, 14, 121, 171, 173, 177
as text and as discourse, 10–12, 103
Pollack, William (author and consultant), 21, 25, 177, 182, 188
"positive" treatment of boys, 51, **71**, 99, 106, 110–11, 112, 113, 114, 122, *see also* negative portrayal of or approach to boys

poverty, 53, 78, 102, 111, 205
power, critique of using term, 119–20
Pride and Prejudice program, 202
private schools
   in Australian history, 35–6
   competition with public schools, 147–8
   funding, 35–6, 158, 173
   proportion of all Australian schools, 35
   in the United States, 186–7
   using boys' education to advertize, 36, 168
   see also public schools
products and services, boys' education, 107, 109, 136, 142, 203, 204
(pro)feminists, 38, 42, 50, 54, 65, 78, 86–94, 97, 98, 99, 101, 108, 112, 113, 117, 122, 126, 150–1, 155, 195, 199–200, 203, 204, 205, 213 (note 26)
   failure to correct "mishearings," 90–1
   language and media use, 203
   professional development, 204
   strategic mistakes, 89–93
   successes, 89, 97–8, 108–9, 122–3, 151
   see also feminists; *under individual names*
professional development, *see under* teachers
progressive groups, 24, 78, 87, 129, 188–9
progressive possibilities for working with boys, 53, 89, 92, 174, 176, 178, 193, 194, 197, 198, 201
Promise Keepers, 22, 187
public schools
   competition with private schools, 147–8, 177
   underfunding of, 173, 190, 204
   in the United States, 187
   see also private schools
publishing, political economy of, 25

qualitative research, *see under* research
quantitative research, *see under* research
Queensland, 33, 35, 81, 83, **131**, **132**

race, 19, 34, 55, 90, 95, 99, 123, 151, 175, 183–4, 190, 194, 202, 205
racism, 99, 102, 111, 121, 194
Reading Recovery, 25, 82
recontextualization, 32, 168, 188
recuperative masculinity politics, 50, 202,
   see also backlashes against feminism
Rees, James (Committee secretariat), 62–3

rejection, 88
relationships, importance of for boys, 71, 79, 80, 84, 92, 128, 198
religious conservatism, 177, 190–1
religious education
   in Australian history, 35
   at Springtown Religious School, 159–60
religious intolerance, 121
research
   ascription of legitimacy by the Committee, 94–5, 105
   attempts by conservatives to limit, 100
   borrowing from other countries, 25, 32
   government control of, 190, 193
   practitioner, on boys' education, 127, 139, 141, 151, *see also* action research
   produced for DEST, 80, 84, 115, 127, **137**, 218 (note 51), 220 (note 49)
   qualitative, Committee's criticisms of, 84, 90, 100
   quantitative, Committee's preference for over qualitative, 83, 85, 90, 100, 113
   as "sexy" or mysterious, 24–5
   translated for teachers, 116, 168, 171
resistance, forms of (Debra Hayes), 88–9,
   see accommodation; rejection; transformation
risk-taking behavior, 2, 19, 109, 193
rites of passage, 182
Riverside Schools Cooperative (RSC), 146–56, 157, 169, 173, 176, 201
Roberts, Ian (rugby player), 42
Rock and Water Program, 107
role models, *see* male teachers and role models
Rowe, Ken and Katherine, 74, 83–4, 85, 94, 101
rurality, 34, 78, 102, 123
   and educational failure, 78
   see also under Australia
Ryan, Susan (Australian senator), 47

Sarra, Chris (principal), 108, 109
Sawford, Rod (deputy chair of Committee), 25–6, 42, 61, 65, 66, **76**, 80, 84, 85, 86, 89, 90, 91, 100, 113
scholarships for male teachers, 2, 7, **8**, 12, 100, 123–6, **137**, 138, **207–8**
school culture, 146, 159–61, 166, 167, 168, 175, 176
   boys' education as threat to, 160–1

schools and teachers as primarily responsible for changing boys, 111–12, 120–1
Schools in Australia, see Karmel Report
Sex Discrimination Act, 7, 125, 176
  attempts to revise, 7, 53, 54, 125–6, **208, 209**
sexism, 2, 99, 118, 120, 121, 198
sexuality, elision of, 98, 185
shootings, school, 19, 21
Sidebottom, Sid (Committee member), 10, 63–4, 66
single parents, 5, 20, 78, 98
single-sex schooling, 6, 12, 35–6, **71**, 83, 98, 159, 186, 187, 189, 191, 193–4
site-based management, 11, 116, 183, 188
situated strategizing, 197, 202–5
Slade, Malcolm, *see* Trent, Faith, and Malcolm Slade
social construction of gender, 48, 90, 91, 98, 117–18, 119–20, 122, 153, 185
social justice, 54, 67, 99, 187, 198, 199, 202, 203, *see also* progressive possibilities for working with boys
socioeconomic status, role of, 83, 86, 90, 98, 114, 123, 175, 183–4
Sommers, Christina Hoff, xii, 21, 26, 177, 182, 183
South Australia, 33, **131–2, 132**
special education, boys' overrepresentation in, 184
Spellings, Margaret (U.S. secretary of education), 192
sport, 29, 36, 40, 42, 56, 159, 167, 181, 191, 204
Springtown Religious School, 12, 146, 156–69, 170, 172–3, 175, 176, 200–1
standardized tests, *see* tests and testing
standards, 11, 18, 138, 187, 193
state as gendered, the, xii, 190, 224 (note 40)
state-Commonwealth tensions, *see under* Commonwealth Government
states and territories, rivalries among, 129
state schools, *see* public schools
steering at a distance, 138, 140–1, 171
strategy, xiv, 194–5, 197, 202–5, *see also* situated strategizing
structures of feeling, 104, 112
submissions to the Inquiry, *see under* Inquiry into the Education of Boys
Success for Boys (SFB), 103, 135, 136, **137**, 138, 198, 200, **208, 209**

suicide, 2, 19, 51, 53, 97, 99, 184
suspensions and expulsions, 5, 19, 184, 205

Tallebudgera Beach School, **76**, 84, *see also* 3R program
tasks of the scholar activist, xiv
Tasmania, 33, **132**, 202
teachers
  burnout and overwork, 79, 164, 176, 199
  concern for boys, 91, 171–2
  education or training, 2, 6, **8**, 47, 66, 68, 83, 98, 100, 123, 124, **208**
  female, criticisms of, 3, 20, 172, 181, 199
  importance of, 83
  knowledge of boys' education issues and research, 170
  males, *see* male teachers and role models
  pay, 7, **8**, 78, 79, 92, 97, 123, 124, 125, 173, 185
  as primarily responsible for changing boys, 111–12, 120–1
  professional development, **8**, 83, 84, 103, 107, 119, 123, 129, 135, 140, 149, 159, 172, 173, 204, *see also* Rock and Water; Success for Boys
  professional discretion and critical thinking, 200–1
  reform fatigue, 175
  as researchers, *see* action research; research, practitioner
  unions, *see* unions; Australian Education Union
teaching methods, *see* pedagogy
technology, *see* information and communication technologies
tests and testing, 5, **7–8**, 18, 22, 54, 67, **71**, 81, 82, 83, 96, 97, 100, 110, 119, 120, 177, 182, 183–4, 187, 190, 193
threshold knowledge, 62, 66, 156
"thrill of the new," 24–5
"tips for teachers," 20, 98, 171
Title IX (U.S. law), 181, 186, 189, 191, 193–4
  reviews of, 191, 193–4
transformation, 89
Trent, Faith, and Malcolm Slade, **76**, 84, 85
triage, educational, 22, 168, 190
typical vs. telling cases, 145

unemployment, 2, 19, 81
uniforms, 157, 167, 182
unions, teachers, 42, 50, 65, **71**, 78, 79, 85, 89, 90, 100, 139, *see also* Australian Education Union
United Kingdom, 22, 32, 142, 187, *see also* England
United States, 2, 3, 4, 11, 17, 19, 20, 21, 22, 23, 24, 25, 26, 32, 34, 35, 37, 41, 43, 44, 52, 55–6, 59, 60, 87, 100, 101–2, 112, 136, 139, 142–3, 158, 169, 176–8, 179–95, 197, 199, 200, 201, 203
  boy panics and crises in, 3
  cultural ambivalence over masculinity, 188
  Department of Education, 191–2, 194
  educational history and policy, 11–12, 180–3
  educational structures, 185–7
  governmental and policy structures, 185–7
  growth trends in boys' education, 189–93
  increasing federal role in education, 190
  policy ecology, educational, 185–8
  politicians concerned about boys, 192–3
  population, 187
  structural limitations on boys' policy, 185–8
  see also No Child Left Behind Act
University of Newcastle, Family Action Centre, 104, 105, 110

"Values for Australian Schooling," 116
victimhood, ascription of and fights over, 51, 67, 99, 192, 193, 194, 197
Victoria, 33, **130**, **132**
violence, 2, 18, 19, 21, 22, 24, 37, 41, 43, 51, 97, 99–100, 111, 120, 192, 198, 201, 202, 205

violence prevention programs, 78, 92, 191, 193, 204
vouchers, 140, 193–4, 203, *see also* choice, school

Western Australia, 33, **131**, **132**
"what about the boys?," xi, 2, 17, 20, 23, 48, 49, 82, 87, 90, 96–7, 111, 138, 142, 155, 159
"what works," 100, 111–12, 139
"which boys?" and/or "which girls?," 2, 20, 42, 88, 89, 90, 98, 115, 123, 146, 151, 152, 155, 185, 204, 205
  as strategy, 146, 151, 155
white victimhood, 194, 197
Whitlam, Gough (former prime minister), 32, 45, 79
whole language reading instruction, **71**, 100
whole-school and -child solutions, 98, 128, 176
women
  in the curriculum, 182
  domestic labor, 37
  earnings, 19, 184
  economic and social standing, 2, 37, 184, 205
  in the government bureaucracy, *see* femocrats
  in higher education, 184
  outcomes of schooling for, 205
  policy infrastructure for, rollbacks of, 138–9, 191, 193
  as politicians, 37, 45
  teachers, *see under* teachers, female
  violence against, 37
Women's Educational Equity Act (U.S.), 181, 191
workplace skills and attitudes, boys', 23, **71**, 80
World War I, 40–1
writing, boys', 149, 152, 153–4, 155, 183

GPSR Compliance

The European Union's (EU) General Product Safety Regulation (GPSR) is a set of rules that requires consumer products to be safe and our obligations to ensure this.

If you have any concerns about our products, you can contact us on

ProductSafety@springernature.com

In case Publisher is established outside the EU, the EU authorized representative is:

Springer Nature Customer Service Center GmbH
Europaplatz 3
69115 Heidelberg, Germany

www.ingramcontent.com/pod-product-compliance
Lightning Source LLC
LaVergne TN
LVHW011809060526
838200LV00053B/3716